CULTURAL INTELLIGENCE & LEADERSHIP

CULTURAL INTELLIGENCE AND LEADERSHIP:

AN INTRODUCTION FOR CANADIAN FORCES LEADERS

EDITOR
Karen D. Davis

CANADIAN DEFENCE ACADEMY PRESS

Canadian Defence Academy Press
PO Box 17000 Stn Forces
Kingston, Ontario K7K 7B4

Produced for the Canadian Defence Academy Press
by 17 Wing Winnipeg Publishing Office.
WPO30408

Library and Archives Canada Cataloguing in Publication

Cultural intelligence and leadership : an introduction for Canadian Forces leaders / edited by Karen D. Davis.

Produced for the Canadian Defence Academy Press by 17 Wing Winnipeg Publishing Office.
Issued by: Canadian Defence Academy.
Includes bibliographical references and index.
ISBN 978-1-100-11464-4 (bound).--ISBN 978-1-100-11465-1 (pbk.)
Cat. no.: D2-238/1-2009E (bound).--Cat. no.: D2-238/2-2009E (pbk.)

1. Military art and science--Canada. 2. Leadership. 3. Cultural awareness. 4. Social intelligence. I. Davis, Karen D. (Karen Dianne), 1956- II. Canadian Defence Academy III. Canada. Canadian Armed Forces. Wing, 17

UB210 C84 2008 355.4 C2009-980028-4

Printed in Canada.

1 3 5 7 9 10 8 6 4 2

ACKNOWLEDGEMENTS

Karen D. Davis and her co-authors, Major Brent Beardsley, Dr. L.W. (Bill) Bentley, Dr. Daniel Lagacé-Roy, Dr. Robert W. Walker, and Justin C. Wright would like to thank Commander (retired) Robert S. Edwards, Craig Leslie Mantle, and Chief Petty Officer 2nd Class Paul Pellerin for their numerous and invaluable contributions to the development of the volume.

Appreciation is also extended to Captain Imam Suleyman Demiray, Dr. Kimberly-Anne Ford, Denise Kerr, Major S. K. Moore, and Dr. Emily Spencer. Your expertise and experience have enriched this volume.

The last word goes to those who worked diligently behind the scenes to ensure that all of our efforts became a professional product. CFLI project manager, Greg Moore, provided numerous suggestions, guidance and attention to important details throughout the publication process. Captain Phil Dawes, Evelyn Falk and Adrienne Popke at 17 Wing Publishing are commended for the technical and creative expertise that created the final polished product.

To each of you, sincere thanks for your contributions.

TABLE OF CONTENTS

LIST OF FIGURES AND TABLES

FOREWORD

The importance of leveraging cultural knowledge to increase operational effectiveness in defence and security operations is certainly not new. In recent decades, however, the complexity of military operations has intensified. This complexity is due, at least in part, to the increasing presence of a virtually indefinite population of international government and non-government organizations, mercenary soldiers, police and paramilitary organizations, coalition forces, and a range of often unpredictable host-nation stakeholders and belligerents. Regardless of the type of mission or where it must be conducted, the security environment is essentially a dynamic myriad of integrated, isolated, and layered human systems, each with their own understandings, motivations, and goals. Consequently, cultural intelligence applies to all interactions within and across domestic and international communities, networks, organizations, militaries, government and non-government organizations, including "whole of government" approaches to security challenges. Effectively negotiating this cultural landscape requires an imposing level of analysis, expertise, and intellectual resource at the policy, strategy, operational, and tactical levels.

Cultural intelligence is an inclusive meta-competency that derives optimum analytical and intuitive power from the integration of knowledge, cognitive, motivation, and behaviour, thus providing a framework for continuous development of the necessary capacity to meet intra- and inter-cultural challenges in the defence and security environment. As such, this volume emphasizes the importance of cultural intelligence as a critical area of development for leaders in the Canadian Forces (CF).

In addition, leaders have a responsibility to lead in a manner that facilitates learning, adaptability, and effective problem solving within their organizations, units and teams. Leaders play a critical role in shaping and influencing the climate and culture that supports effective leadership, including the development and application of cultural intelligence. For example, an open culture in which people are encouraged to engage in inquiry, think critically, think differently, and discuss varying perspectives and possible explanations of unfamiliar activities and behaviours, provides a fertile environment for the growth of cross-cultural capacity. Importantly, an open culture has also been identified in CF doctrine, *Leadership in the Canadian Forces: Doctrine*, as a key condition supporting leadership development.

Cultural Intelligence and Leadership: An Introduction for Canadian Forces Leaders is a key resource for the professional development of all military leaders as well as a valuable tool for generating dialogue and further development of multicultural and "whole of government" capacity across the CF. I urge you to read it!

Major-General Daniel Gosselin
Commander, Canadian Defence Academy

INTRODUCTION

Military leaders have been applying cultural intelligence (CQ) for centuries; however, not all have been as successful as others. CQ is the ability to recognize the shared beliefs, values, attitudes and behaviours of a group of people and, most importantly, to effectively apply this knowledge toward a specific goal[1] or range of activities.The abbreviation "CQ" is used to reflect cultural intelligence as a facet of intelligence,[2] resulting from the integration of several human dimensions to facilitate the success of leaders within national, international, host-nation, and enemy domains,[3] As a framework for understanding what contributes to mission success across culture, CQ also provides a road map to inform the development of the relevant capacities as an integral component of professional competence.

History is replete with examples of how cultural intelligence has, and has not, been applied in warfare. Interestingly, in almost each battle recounted in the Peloponnesian War, the three decade-long fight between the Spartans and the Athenians at the end of the fifth century B.C., Thuycidides described how each side took advantage – or tried to – of the other's cultural "weaknesses", while at the same time enhancing their own cultural "strengths."[4] We live in a somewhat different world today, a global community that is more complex and comprised of an endless number of groups, organizations, societies, nations, and belief systems that are similar and different in many ways. As a result, culture and its impact on military operations are receiving heightened attention.

It is widely recognized that CQ among all military leaders is an important contributor to mission success, whether that means combat superiority or working effectively with internal subcultures, external domestic and international organizations, or other military organizations. There is also a renewed emphasis on effective leaders in the CF exemplified by the publication of the suite of CF leadership manuals in the last few years. These manuals amplify the CF definition of leadership at both the level of leading people and leading the institution. CF leadership is defined therein as: "directing, motivating and enabling CF personnel to accomplish their tasks while developing and improving capabilities to ensure mission success." CQ is one such vital capability and it is the responsibility of the senior leadership of the CF to make sure that it is properly understood, codified and imparted to all members of the profession of arms through the professional development system. In today's complex security environment, we cannot afford to leave CQ to chance.

Much of the renewed focus on culture and CQ is a direct result of Canada's experience in Afghanistan; however, the development and application of CQ has contributed to a number of important military successes and failures in the past. Although not new, the military increasingly views culture and CQ as priorities that must be integrated with other operational imperatives at the strategic, operational and tactical levels of operations. CQ is an essential contributor to the ability to determine adversarial intent; to work effectively across joint, inter-agency, multinational and public domains (JIMP), to access and exercise whole of government (WoG) approaches, and to negotiate the demands of interrelated defence, diplomacy, and development (3D) objectives.

All leaders in the CF, commissioned and non-commissioned alike, require CQ to do their jobs. It is an integral aspect of leader development, to which many of the capacities and attributes of effective leadership contribute. However, leadership development, without particular attention to those competencies that contribute to effective performance across organizations, ethnic communities, and societies worldwide, is not enough to achieve effective CQ.

A key objective of this volume is to contribute to the professional development of CQ capacity across the CF by laying a foundation upon which further discussion, exploration, discovery, and development can take place. It will be particularly useful to those who have a responsibility for the development of effective military teams, as well as the application of effective tactics and strategies in domestic and international operational environments. It is also intended as a practical resource upon which to build further understanding and competency in the CQ domain. As such, the chapters are comprised of discussions of the theory and concepts necessary to frame one's understanding of the context within which CQ is relevant, what it is, why it is important, and how it can be developed and applied within the CF. Every effort has been made to integrate and streamline the concepts and theories that contribute to effective CQ in the CF; however, as a critical enabler of mission success it is imperative that the discussion continue. The development of CQ across the CF is dependent upon reflection, dialogue, continuous learning, lessons learned, and most importantly, the application of adaptable and innovative critical analyses to cultural challenges and dilemmas.

Chapter 1, "Systems Theory, Systems Thinking and Culture," describes the context within which CQ is a critical enabler of mission success. Culture is an integral part of any human activity system be it a civilization, a nation, any organized social or political community, or any large institution or organization such as the church or a business enterprise. These systems are non-linear and must be conceived holistically. Culture, and indeed, CQ, can only be fully understood by employing systems thinking. This first chapter introduces systems theory and systems thinking in reference to culture, in order to provide the foundation for a more detailed and thorough discussion of culture and CQ in the chapters that follow.

"Culture and Cultural Intelligence," are discussed in relative detail in Chapter 2, which provides an introduction to the concept of culture and the multi-dimensional character of CQ, approaches to understanding culture at societal and organizational levels, and the relevance of CQ to mission success. Importantly, this chapter emphasizes the inherent complexity and challenge that culture presents, as well as the potential power of CQ that results from the integration of knowledge, cognition, motivation, and behaviour.

The successful development and execution of Canada's national security policy is dependent upon a thorough familiarity with Canadian national strategic culture, as well as some knowledge of the strategic culture(s) of adversaries. As such, strategic culture is an important element of CQ. Chapter 3, "Cultural Intelligence and Strategic Culture," places the concept of CQ firmly within the security domain through a discussion of the concept of strategic culture and its contributing factors. An overview of Canada's strategic culture is used to illustrate the type of analysis that contributes to an enhanced understanding of the strategic culture of a nation or political community, and thus to the overall CQ capacity of the CF.

Chapter 4, "Cultural Foundations," emphasizes that although CQ is a critical competency for the success of individual CF members and the overall capacity of the CF when operating in unfamiliar environments, its development begins at home within relatively familiar domains. The chapter highlights the ways in which our cultural orientations contribute to, and have the potential to impair and enhance, our CQ. Key challenges in this regard include developing an awareness of the strengths and biases of one's own culture, and recognizing when and how our own cultural orientations are influencing our beliefs and behaviours, as well as our ability to interpret and understand the beliefs and behaviours of others. The way in which others perceive the CF and its members within various contexts is also presented as an important aspect of understanding one's self and own culture.

Building upon notions of identity and cultural foundations, Chapter 5, "Cultural Intelligence and Identity Development," establishes the relationship between identity development, moral reasoning, and CQ. Based upon these relationships, the chapter employs the five stages of Robert Kegan's identity development framework as the foundation upon which the process for *making sense* of one's meaning of culture is based. There is always a balance to strike between cultural relativism and imposing one's own values on an unfamiliar situation. This chapter illustrates how the capacity to make such decisions is dependent upon the interplay between the extent to which one's own identity reflects an internalization of professional and moral values, and the level of CQ that has been achieved. A case study of the first CF Muslim chaplain deployed to Afghanistan (presented at Appendix A), provides a practical example of the successful application of CQ to religious identity in a host-nation operational domain.

The effective application of CQ is, of course, dependent upon the extent to which CQ capacity has been developed among CF members. Chapter 6, "Developing Cultural Intelligence," provides a strategy for that development within the broader context of the CF Professional Development Framework (PDF). The chapter highlights the relevance of a variety of learning processes to the development of CQ, including experience, exposure, focused assignments, and foreign engagements, which complement other adult-learning and professional development strategies. Particular emphasis is placed on the critical role of andragogy in contributing to the development of CQ within the PDF. A case study of the impact of single-service cultures on CF unification, presented in Appendix B, provides a compelling argument for the need for professional development (PD) and CQ to achieve mission success, not only on the international and inter-organizational scene, but also in the presence of internal and domestic intra-institutional challenges.

Chapter 7, "Applying Cultural Intelligence in the Canadian Forces," provides a conceptual framework for the employment of CQ by CF leaders at the tactical, operational, and strategic levels. In addition, the application of CQ by advisors at the policy level is considered. Utilizing the general system of war and conflict as a framework, the discussion underlines the imperative of integrating CQ throughout all planning and execution tasks at national, international, host-nation, and enemy domains to achieve mission success. This discussion highlights the requirement to apply non-linear methodologies, suitable for addressing significant levels of complexity, to the development of policy, strategic direction, cultural overlays, and campaign planning. A case study of the CF and the 1991 "Oka Crisis" included at Appendix C illustrates the potential pay-offs of ensuring that CQ is integrated across all levels – from tactical to policy.

The ways in which peoples from different nations (or other communities and groups) interact take on various forms, from conflict to co-operation to cohabitation. Relations among peoples can be described as interactions between distinct, complex systems; distinct in that there are two or more distinguishable groups, containing members who identify with the group's values, beliefs and norms – with the group's culture. These international or intercultural relations are situated within a larger complex system – humanity. Coming to an advantageous understanding of these complexities within the context of security and conflict operations is challenging. This volume serves as a caution against underestimating the impact of culture or relying upon culture-specific knowledge and observable behaviours to guide strategy, planning, decisions and action in both international and domestic arenas. Perhaps, most importantly, this volume is a first step toward opening the dialogue and learning that will contribute to a greater CQ capacity across the CF.

NOTES

1 Emily Spencer, *Crucible of Success: Applying the Four CQ Domain Paradigm, The Canadian Forces in Afghanistan as a Case Study*, (Kingston, ON: Canadian Forces Leadership Institute, 2007), 3. This definition was expanded by a Canadian Forces Leadership Institute CQ project team to include "or a range of activities."

2 Although CQ does not represent a mathematical relationship of capabilities in the same way as an intelligence quotient (IQ), it is considered to be a unique facet of intelligence; the usage of CQ parallels the use of Emotional Quotient (EQ) to refer to emotional intelligence. See Christoper Earley and Soon Ang, *Cultural Intelligence: Individual Interactions Across Cultures*, (Stanford, California: Stanford University Press, 2003), 4.

3 For discussion of these domains as a potential paradigm for CQ, see Emily Spencer, *Crucible of Success*.

4 Ibid., 9.

CHAPTER 1

SYSTEMS THEORY, SYSTEMS THINKING AND CULTURE

Dr. Bill Bentley

Culture is a somewhat elusive concept and many different definitions can be found in the social science literature, especially anthropology. As one noted social scientist has observed: "In contrast with social structure, the concept "culture" is one of the spongiest words in social science, although, perhaps for that reason, in the hands of an expert, enormously useful."[1] Significantly, one common feature of most of the better definitions is that culture is depicted as a system. For example, the anthropologist Clifford Geertz states that "cultural assumptions constitute a system of inherited conceptions expressed in symbolic forms by means of which men communicate, perpetuate and develop their knowledge about, and attitudes toward, life."[2] This first chapter will introduce systems theory and systems thinking in reference to culture, in order to provide the foundation for a more detailed and thorough discussion of culture and CQ in the chapters that follow.

CULTURAL SYSTEMS AND SYSTEMS THEORY

A leading authority on organizational culture and leadership, Edgar Schein, defines culture thusly:

> "A system of shared basic assumptions that a group has learned as it solved its problems of external adaptation and internal integration, that has worked well enough to be considered valid and, therefore, to be taught to new members as the correct way to perceive, think, and feel in relation to their problems."[3]

In the security field, "strategic culture," described in detail in Chapter 3, has been defined as a system of symbols comprising two parts; the first consists of basic assumptions about the orderliness of the strategic environment, that is about the role of war in human affairs; and second, about the efficacy of the use of force.[4]

Culture as a system is usually analyzed by identifying the sub-systems listed below and determining how they interrelate and interact:

- Observed behavioural regularities when people interact;
- Group norms;
- Espoused values;
- Formal philosophy;
- Rules of the game;

- General feeling in the group (climate);
- Embedded skills;
- Habits of thinking – mental models;
- Shared meanings; and
- "Root metaphors" or integrating symbols.

To make the matter even more complex, culture is embedded in a given society. Culture and society are inextricably intertwined and must be considered holistically.

A society can be defined as a population whose members are subject to the same political authority, occupy a common territory, have a common culture and share a sense of identity.[5] No society is homogeneous, however. A society usually has a dominant culture, but can also have a vast number of secondary cultures. Societies are not static; they change over time. Each society is composed of both social structure and culture. Social structure refers to the relations among groups of persons within a society. It is persistent over time, and regular and continuous despite disturbances. In a society, the social structure includes groups, institutions, organizations and networks. Social structure also involves the following:

- Arrangement of the parts that constitute society;
- Organization of social positions; and
- Distribution of persons within those positions.

Culture, associated with such a social structure can be thought of as a "web of meaning" shared by members of a particular society or group within society. Culture is, therefore:

- *a system* of shared beliefs, values, customs, behaviours and artifacts that members of a society use to cope with their world and one another;
- learned through a process called enculturation;
- shared by members of a society (there is no *culture of one*);
- patterned, meaning that people in a society live and think in ways forming definite, repeating patterns;
- changeable through social interaction between people and groups;
- arbitrary, meaning assumptions cannot be made regarding what a society considers right and wrong, good or bad; and

- internalized, in the sense that it is habitual, taken for granted, and perceived as natural by people within a society.

It is, therefore, crucial to view the concept of culture through the lens of systems theory. Understanding culture, and indeed, CQ, can only be fully achieved by employing systems thinking.

The beginnings of the systems movement and the development of systems theory can be traced to the turn of the 19th century and the emergence of Romanticism and Romantic science, especially biology. However, the explicit debate about how to deal scientifically with non-linear, complex systems began in earnest in the second half of that century. Some of the first tentative expositions of what later became known as systems thinking were written in the 1920s. A pioneer in systems theory and the founder of general systems theory, Ludwig von Bertalanffy, wrote extensively on the subject in the 1940s and in 1954 he helped found the Society for General Systems Research.

To what kind of systems is systems thinking applied? The "systems map," developed by British scientist and management consultant Peter Checkland, and illustrated in Figure 1-1 below, is a helpful way of conceptualizing the nature of different systems. A brief description and comparison of Checkland's three major systems follows.

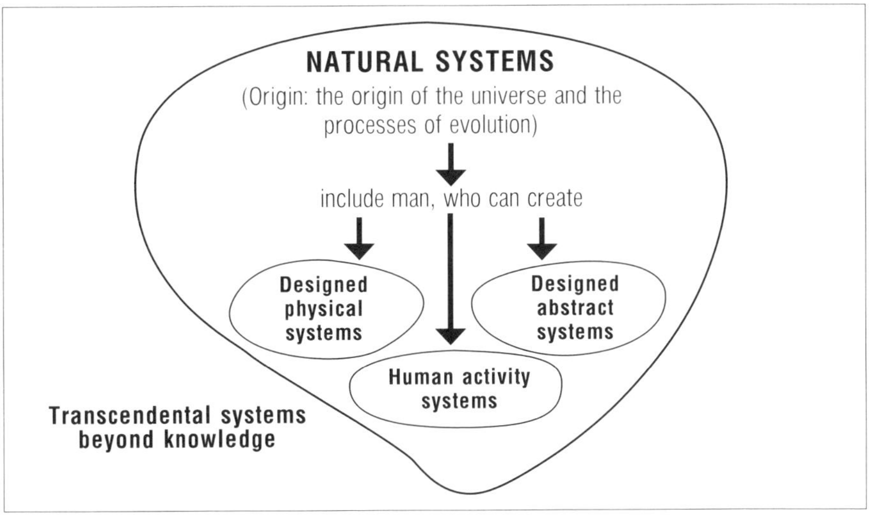

FIGURE 1-1: SYSTEMS MAP[6]

Designed physical systems such as a national energy grid or a missile defence system are amenable to "hard systems" thinking. Examples of methodologies to deal with these kinds of systems include systems engineering, operational research (OR), RAND* systems analysis, the viable system model and systems dynamics. All these approaches are positivistic, making the assumption that the system of concern can be named and can be manipulated in the interest of efficiency. None of these approaches pays attention to the existence of conflicting world views, something which characterizes all social and cultural interaction. Designed abstract systems such as mathematical systems or philosophical systems are also amenable to hard systems thinking and analytical reasoning. Human activity systems are, however, very different. Such systems are complex, interactive and non-linear and are only effectively dealt with using "soft systems" thinking. Marine Lieutenant-General Paul Van Riper describes physical and abstract systems as structurally complicated, whereas, human activity systems are interactively complex.[7]

Modern soft systems thinking is strongly influenced by phenomenology, hermeneutics and critical theory. One of the main ideas in phenomenology is that of intentionality – the concept that all conscious mental activity is thinking about something. Phenomenology's main concern is the nature and content of our thinking about the world, rather than the world itself as something independent of all observers (the world of positivism). Hermeneutics is the study of meaning – the whole philosophical process is one of seeing, clarifying, determining and distinguishing meanings. Hermeneutics' chief concern is to establish that the subject matter of the human sciences is intrinsically different from that of the natural sciences, being concerned not with external facts but with expressions of the human mind which become cultural artifacts by a process of "objectification." Hermeneutics comprises a circular process of discovery called "the hermeneutic circle," a means of perceiving social wholes as both wholes and parts. Critical theory, derived from the work of the so-called Frankfurt School combines an attack on positivism as an adequate base for social science with a belief that adequate social theory must perceive society as a set of processes which emancipated man can change. A leading theorist in critical theory, Jürgen Habermas, contrasts purposive, rational action governed by technical rules based on empirical knowledge, and communicative action, or symbolic interaction, governed by binding consensual norms that define reciprocal expectations about behaviour. The former action involves linear, analytical thinking, whereas the latter involves systems thinking.

Clearly, culture is an integral part of any human activity system, be it a civilization, a nation, any organized social or political community, or any large institution or organization such as the church or a business enterprise. These systems are non-linear and must be conceived holistically.

---

* The RAND Corporation is an independent research organization.

COMPLEX, NON-LINEAR SYSTEMS

For a system to be linear, it must meet two simple conditions. The first is proportionality, indicating that changes in system output are proportionate to system input. Such systems display what in economics is called "constant returns to scale," implying that small causes produce small effects and large causes generate large effects. The second condition of linearity, called "additivity or superposition," underlies the process of analysis. The central concept is that the whole is equal to the sum of the parts. This allows the problem to be broken up into smaller pieces that, once solved, can be added back together to obtain the solution to the original problem.

Non-linear systems are those that disobey proportionality and additivity. Interactively complex, non-linear systems are highly sensitive to inputs: immeasurably small inputs can generate disproportionately large effects. In addition, with interactive complexity it is often impossible to isolate individual causes and their effects, since the parts are all connected in a complex web. Interactive complexity produces fundamentally unpredictable and even counter-intuitive behaviour.

A complex system has the following characteristics:

- it involves large numbers of interacting elements;
- the interactions are non-linear, and minor changes can produce disproportionately major consequences;
- the system is dynamic, the whole is greater than the sum of the parts, and solutions can't be imposed (rather, they arise from the circumstances);
- the system has a history, and the past is integrated with the present, the elements evolve with one another and with the environment, and evolution is irreversible;
- though a complex system may, in retrospect, appear to be ordered and predictable, hindsight does not lead to foresight because the external conditions and systems constantly change; and
- in a complex system the agents and the system constrain one another, unlike in ordered systems in which the system constrains the agents, or chaotic systems in which there are no constraints.

Non-linear systems such as culture are sometimes called complex adaptive systems. It is here that complexity theory and chaos theory come to bear. The general model of such organized complexity is that there exists a hierarchy of levels of organization, each more complex than the one below. Thus, in any culture one finds explicit, espoused values and then above that are deeper, implicit values. The higher levels are characterized by emergent properties that do not exist at the lower level. Moreover, these emergent properties are meaningless at the lower level.

Hierarchy theory is concerned with the fundamental differences between one level of complexity and another. Its ultimate aim is to provide both an account of the relationship between different levels and an account of how observed hierarchies come to be formed, what generates the levels, what separates them and what links them.

In a hierarchy of systems, maintenance of the hierarchy will entail a set of processes for purposes of regulation or control. All these control processes depend upon communication: a flow of information, be it manual or automatic, in the form of instructions or constraints.

SYSTEMS THINKING

Thinking about such complex, non-linear systems is in contrast to the analytical thinking that underlies conventional planning procedures. It is important to note that analytical thinking is one of the key attributes of CF professional development (presented in Chapter 6), and as such is complementary in many ways to systems thinking. However, it is also important to understand the limitations of analytical thinking within the context of complex systems. Analytical thinking decomposes a subject successively into parts until it can explain the behaviour of each of the separate parts and then seeks to explain the whole as an aggregation of the behaviour of the parts. This process is sometimes called reductionist. In contrast, systems thinking places the system in the context of the larger environment of which it is a part and studies the role it plays in the larger whole. Systems thinking is, thus, expansionist, progressing outwards to consider broader and broader systems. Systems thinking can be compared to analytical thinking in the following way. Systems thinking is:

- holistic rather than reductionist;
- interpretation rather than analysis;
- design rather than planning;
- re-understanding rather than re-engineering;
- subjectivism rather than objectivism;

- conceptualizing rather than universalizing; and
- critical reflection rather than identification.

The sub-systems of a cultural system, enumerated above, cannot be analyzed in a linear manner, adding one to the other until arriving at the *whole*. They are interactive, producing emergent phenomena and, therefore, must be viewed holistically. They need to be interpreted using the methods of phenomenology and hermeneutics, not analyzed like the engine of a car. The cultural observer must, in some sense or another, *feel* the culture in a subjective sense, not read it as a set of empirical data. It must be conceptualized in its uniqueness, not generalized to identify universal ideal-types or common rules and structures.

There are a number of approaches or methodologies that have been developed to employ systems thinking in a practical manner. Soft systems methodology (SSM), developed by Peter Checkland, is one of the best that has been developed over the past 30 years.[8] It has been applied in hundreds of cases in both the public and private sectors. SSM is an organized, flexible process for dealing with complex adaptive systems that require understanding and continuous learning. The "process" referred to is an organized process of thinking your way through to making sensible judgements about a human situation and it is a process based on a particular body of ideas, namely systems ideas.

The core concept within SSM is, therefore, that of an adaptive whole which can survive through time by adapting to changes in its environment. Clearly, it is another way of looking at culture as defined at the beginning of this chapter.

SUMMARY

The application of a tool such as SSM assumes that, as a minimum, its operators recognize that culture operates through and within complex systems, the boundaries of which are limited only as defined for the purpose of analysis and interpretation. That is, as highlighted in this chapter, systems are expansionist and have limitless potential to influence other systems. In addition, successful application of SSM is dependent upon the level of CQ that informs the methodology. That is, the extent to which cultural knowledge, cognition, motivation and behaviours have been developed as an integrated meta-competency that we call CQ, will inform the utility of SSM and other approaches to interpreting and negotiating the complex relationships among cultures and systems. Chapter 2 provides further explanation of the concept of CQ, as well as its relationship to culture and complex systems.

NOTES

1 C. Wright Mills, *The Sociological Imagination*, (Oxford: Oxford University Press, 1959), 160.

2 Clifford Geertz, *The Integration of Cultures*, (New York: Basic Books, 1973), 89.

3 Edgar Schein, *Organizational Culture and Leadership*, (San Francisco: Jossey-Bass, 1992), 12.

4 Alastair Johnston, "Thinking About Strategic Culture," *International Security* 19, 4 (1995), 46.

5 United States. *The US Army and Marine Corps Counterinsurgency Field Manual*. (Chicago: The University of Chicago Press, 2007).

6 Peter Checkland, *Systems Thinking, Systems Practice*, (New York: John Wiley & Sons, 1999), 112.

7 Discussion between Lieutenant-General Paul Van Riper and Lieutenant-Colonel Bentley (Ret'd) of the Canadian Forces Leadership Institute, 18 September, 2006.

8 Peter Checkland, *Soft Systems Methodology in Action*, (New York: John Wiley & Sons, 1999).

RECOMMENDED READING

Checkland, Peter. *Soft Systems Methodology in Action*. New York: John Wiley & Sons, 1999.

Checkland, Peter. *Systems Thinking, Systems Practice*. New York: John Wiley & Sons, 1999.

Geertz, Clifford. *The Integration of Cultures*. New York: Basic Books, 1973.

Johnston, Alastair. "Thinking About Strategic Culture," *International Security* 19, (4, 1995) 32-64.

CHAPTER 2

CULTURE AND CULTURAL INTELLIGENCE

Karen D. Davis and Justin C. Wright

CQ is expressed as the ability to recognize the shared beliefs, values, attitudes and behaviours of a group of people and, most importantly, to effectively apply this knowledge toward a specific goal[1] *or range of activities.*Cultural intelligence is commonly misconstrued as being synonymous with cultural knowledge; however, cultural intelligence is much more than simple *knowledge* of another *culture*. The use of the abbreviation CQ emphasizes this distinction. This chapter provides an introduction to ways of understanding CQ through a discussion of culture, the integrated dimensions that comprise CQ, culture and institutions, and the relevance of CQ to mission success.

The notion that there can be such a thing as CQ is based upon the premise that culture can be defined as well as understood. The concept of culture can be simultaneously elusive, all-embracing, and contradictory, to the point that some within the disciplines of sociology and anthropology have discarded the concept altogether.[2] Although the notion of culture is complex, as a concept, it is a useful tool to employ as a framework to understand the ways in which historically transmitted patterns of meanings, symbols, conceptions, values, and knowledge influence the attitudes, motivations, and behaviours of groups of people. The elusive character of culture also highlights the inherent challenge and complexity it presents to an outsider.

Ronald Wright defines culture from a technical anthropological perspective to include "...the whole of any society's knowledge, beliefs, and practices..."[3] Culture, according to Wright, is everything:

> ...from veganism to cannibalism; Beethoven, Botticelli, and body piercing; what you do in the bedroom, the bathroom, and the church of your choice (if your culture allows a choice); and all of technology from the split stone to the split atom. Civilizations are a specific kind of culture: large, complex societies based on the domestication of plants, animals, and human beings. Civilizations vary in their makeup but typically have towns, cities, governments, social classes, and specialized professions.[4]

Recent scholarship, for the most part, is not quite this ambitious or far-reaching. Many cultural projects focus on conceptions of culture at various levels of society, including national, community, and organizational/institutional domains, thus providing relatively well-defined units of analysis to facilitate understanding. However, culture is an intangible entity that permeates boundaries while simultaneously being influenced by entities from both the inside and outside; that is, culture is not fully contained within defined structures and groups. For example, the military, as a national public institution is embedded within the national "cultural" environment.

Culture can also be conceptualized as the link between an individual and the social group, organization or milieu in which he or she lives. Developmental psychologist Robert Kegan asserts that an individual structures his/her identity by *making sense* of their surrounding world.[5] In other words, the commonly shared beliefs and practices of a given social group form the basis of the individual's social identity. In turn, the individual's unique position in the social group, which is made up of his/her biography, experiences, and interpretation (i.e. his/her process of making sense or making meaning) of his/her culture, comes to maintain and sometimes shape the shared values and practices of the social group. In this way, culture remains continuous as it is passed on through generations, and is at the same time flexible as it adapts itself to the shared experiences and interpretations of the collection of individuals of a given social group, in specific moments of history. The direct relevance of meaning-making processes to CQ is discussed in greater detail in Chapter 5.

The reciprocal relationship between the individual and the social group to which he or she belongs, raises another element of culture, specifically, individual agency.[6] Although culture, or commonly accepted and shared values and behaviours, informs how individuals act and ought to act, culture's influence is not deterministic. In other words, culture communicates the accepted values and practices of a social group, but it is still up to the individual to interpret those values and practices. The interaction between cultural values and behaviour and the individual's unique experience within that cultural setting can be conceptualized as that individual's meaning-making process. Because the meaning-making process is specific to the individual, different perspectives and interpretations of the merit of cultural values and the expression of cultural practices emerge. Although cultural values and practices are shared, many are not endorsed by all. Generally speaking, the existence of these differences, which are responsive to the unique experience of individuals, is the vehicle through which the culture of a social group, organization or milieu may adapt and change.

It is clear from the above discussion that understanding a culture is a very significant and complex task. Even grasping the complex nature of one's own culture, as discussed in greater detail in Chapter 4, is daunting when you begin to consider the diversity of social groups, milieu, and organizations; the relationships between these social entities; that many of these social groups have competing or conflicting interests and goals; that the values and practices of cultures and subcultures adapt to the experiences and perspectives of their members; and that all of these cultural and sub-cultural dynamics are intimately tied to the foundational institutions of a given nation-state, such as its economy, government, or its religious organization(s). Given the difficulty of grasping the inner-workings of the culture in which one lives, the notion of trying to *understand another culture* from without seems, at first glance, unrealizable.

Although it is true that studying a culture will always be a work in progress, there are some broad-stroke factors – elements of human experience that are, arguably, universal in character – through which a rough metric may be applied to begin to understand cultural differences. CQ provides the framework for developing the capacity to understand beyond knowing factual knowledge and applying prescriptive behaviours which will inevitably face limitations within dynamic and unfamiliar cultural milieus.

CULTURAL INTELLIGENCE: A MULTI-DIMENSIONAL CAPACITY

CQ, in a military context, has been understood predominantly in terms of knowledge and cultural awareness derived frequently from the analysis of social, political, economic, and other demographic information that provides an understanding of a people or a nation's history, institutions, psychology, beliefs, and behaviour. A range of skills and competencies have been understood to supplement knowledge and cultural awareness, including cultural sensitivity, relationship building, cross-cultural communication, language ability, listening and observations skills, etc. CQ is not any one of these things in isolation; however, it does *encompass* most of these things. For example, skills such as relationship building, analysis of information and tactical decision-making in operational settings are influenced by cognitive and motivational components that comprise CQ.

CQ experts have invariably constructed CQ as a tripartite or multi-dimensional model. For example, Christopher P. Earley and Soon Ang's research focuses on the relationship between the cognitive, behavioural, and motivational dimensions comprising a CQ construct, while other models also include such dimensions as knowledge, skills, and attitudes.[7] Within his conception of CQ, David C. Thomas places particular confidence in the role of *mindfulness*, as a meta-cognitive strategy that functions as a key mediating link between cognition and behaviour. According to Thomas, mindfulness regulates cognitive processing and response by:

- bringing to mind knowledge relevant to the focus of attention;
- choosing not to respond automatically;
- inhibiting undesirable responses; and
- editing responses to be consistent with motives and goals.[8]

Figure 2-1, below, provides an illustration of the predominant relationships and competencies that contribute to CQ.[9] It is important to note that CQ results from the integrated power of this range of dimensions; more specifically, highly effective CQ cannot be achieved in the absence of any of these dimensions. The dimensions and competencies identified in Figure 2-1 are also discussed in further detail below.

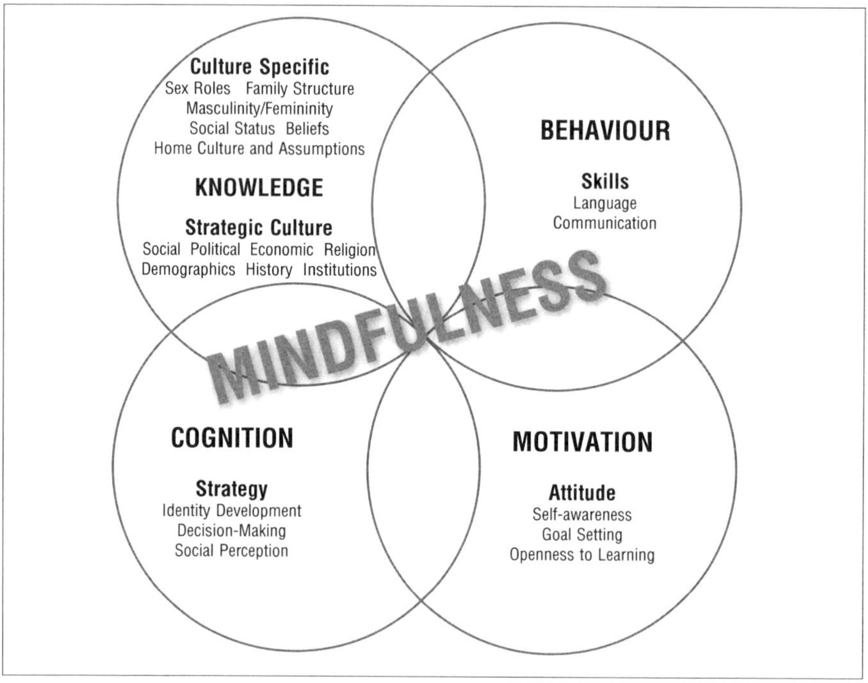

FIGURE 2-1: CULTURAL INTELLIGENCE: CONCEPTS AND RELATIONSHIPS

Importantly, knowledge and cognition are closely related dimensions of CQ. Strategic culture, within the knowledge domain, includes much of what has been commonly understood as intelligence, and sometimes referred to as cultural intelligence in military communities. For example, John P. Coles, Commander, United States Navy, defines cultural intelligence as "analyzed social, political, economic, and other demographic information that provides understanding of a people or nation's history, institutions, psychology, beliefs (such as religion), and behaviors."[10] The importance of strategic culture to CQ, as well as force structure, defence organization, military theory and military doctrine is discussed in greater detail in Chapter 3. Cultural awareness or culture-specific knowledge such as social and family structure, and notions of masculinity and femininity also falls within, but does not wholly comprise the knowledge dimension or fully constitute CQ.

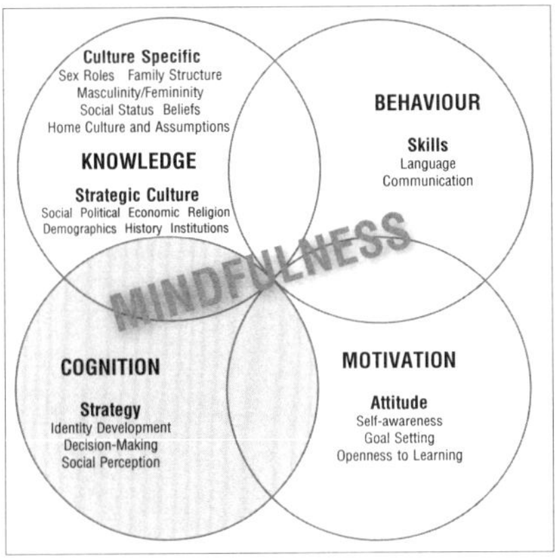

The cognitive scripts that guide behaviour are informed by cultural values which are stored in memory through gradual internalization of prevailing cultural patterns. Specific knowledge such as cultural awareness has the potential to alter cognitive processes; however, prevailing cultural roots continue to influence perception and behaviour.[11] Social perception, within the cognitive sphere, includes, for example, perceptions of events and attributions of their causes; that is, perceptions of what actually took place and why or what led directly to a particular outcome. The relationship between identity development and cultural values systems informs the cognitive processes underlying decision-making and moral reasoning. These relationships are discussed in greater detail in Chapter 5.

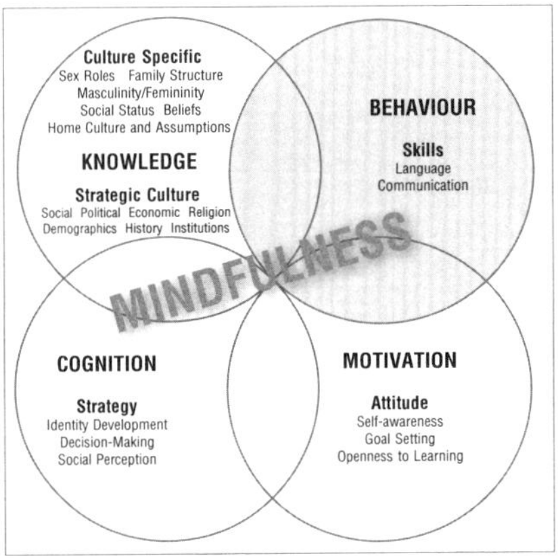

The behaviour/skills dimension of CQ refers to self-presentation based on inputs from the cognitive and motivational dimensions, including, for example, language skills and cross-cultural communication skills. It is important to note that behaviour is based upon understanding, acquired through the cognitive dimension, of what is acceptable or effective within a new culture,[12] such as producing the right tones when attempting to speak a language[13] that one is learning or making decisions in reference to appropriate greetings, handshakes, etc. The capacity to appropriately adapt behaviour to respond to various and dynamic cultural contexts is reliant upon knowledge, cognition and motivational capacities.

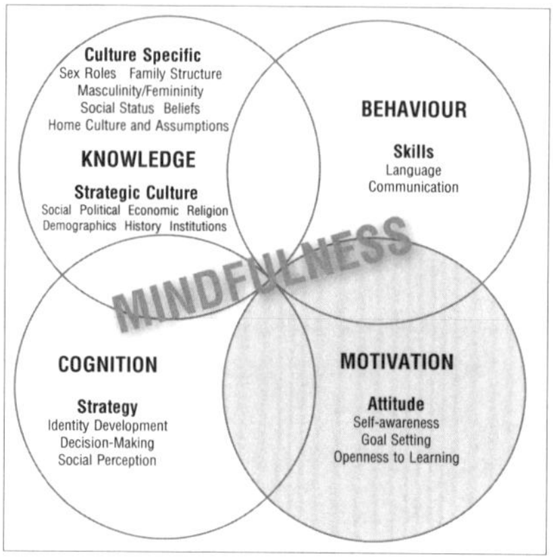

The motivational dimension represents concepts such as awareness of others, self-awareness, perceptual acuity, flexibility, communication, empathy, openness, openness to learning, and goal setting, which facilitate adaptable approaches in culturally unfamiliar or complex situations.[14] A motivated individual would be, for example, someone who strives toward self-awareness and self-development in enhancing their effectiveness in different cultural situations. Within a CQ context, openness to learning reflects a willingness to suspend judgement while accessing new cultural knowledge until a fuller and integrated cultural picture has developed.

Mindfulness, according to Thomas, is a key component linking knowledge and behavioural capability.[15] When developed to a very high level, CQ relies upon the following cognitive activities comprising mindfulness:

- being aware of our own assumptions, ideas and emotions;
- noticing what is apparent about the other person and tuning in to their assumptions;
- using all of the senses to perceive situations;
- viewing the situation from several perspectives, that is with an open mind;
- attending to the context to help understand what is happening;
- creating new mental maps of other peoples' personality and cultural background to assist in responding to them;
- seeking out fresh information to confirm or negate the mental maps; and
- using empathy to understand the situation from another's cultural background.[16]

Mindfulness integrates the unique dimensions of CQ to reflect a *cultural professional*, in much the same way that professional ideology is a key facilitator of the technical, cognitive, social, and change capacities represented in the CF Professional development framework (PDF)[17] discussed in Chapter 6. Mindfulness is presented in Figure 2-1 as a reminder of its role in influencing and synchronizing the knowledge, cognition, motivation, and behavioural dimensions of CQ.

Numerous competencies and attributes that comprise CQ are also complementary to, and at times the same as, those attributes that have been identified as important dimensions of CF leader development. For example, flexibility, communication, and strategic thinking have been identified as important attributes within the PDF, as well as within various CQ models.[18] Reflecting leadership and CF culture, CQ is a meta-competency that has the potential to significantly enhance mission success within culturally different and complex security environments.

Importantly, the following characteristics of CQ inform much of the discussion which is presented in this chapter and this volume overall:

- CQ is a multi-component competency, and in this sense it is distinct from general theories of intelligence, which tend to focus on cognitive capacities only;
- CQ is multi-leveled and can, therefore, be evaluated at various levels;
- CQ, like intelligence in general, is hierarchical, for example, higher order abilities are applicable to strategic planning and selection strategies, and problem solving in specific intercultural interactions are based on lower order abilities;
- High CQ enhances one's ability to adapt;
- High CQ facilitates the shaping and reshaping of one's image; and
- High levels of CQ can be learned.[19]

The multi-component or multi-dimensional nature of CQ is particularly important as it is this aspect that provides the basis for the power of CQ. Although CQ is comprised of unique components, the power of CQ derives from the integration of all the components and dimensions. The link between CQ and mission success is also important to understand.

MISSION SUCCESS AND CULTURAL INTELLIGENCE

CF leadership doctrine[20] defines the effectiveness of the CF in terms of four major outcomes: mission success; external adaptability; internal integration; and member commitment and well-being. These outcomes are influenced both individually and collectively, by military ethos and professional ideology, as illustrated in the Canadian Forces Effectiveness Framework at Figure 2-2, below. The relationship between the four outcomes represents a competing values paradigm, meaning that while each enabling outcome is important in and of itself, the outcomes can also compete with one another for dominance. Mission success is the primary outcome; however, it is dependent upon striking the right balance among all outcomes. As these value domains act in both complementary and competitive ways, the values, symbols, and meanings that have become embedded within CF processes and practices over time – CF culture – contribute to activity and outcome in all domains. The structures, accountabilities, and reporting relationships across the CF contribute to the institutional culture, and also combine to create a system of complex relationships – a complex system that strives to negotiate the right balance across the enabling outcomes to ensure that the primary outcome, mission success, is achieved.

FIGURE 2-2: CANADIAN FORCES EFFECTIVENESS FRAMEWORK

Importantly, there is a reciprocal relationship between the CF Effectiveness Framework and CQ. CQ contributes to mission success and external adaptability through effective outcome at the enemy, host-nation, and international domains. Internal integration, member well-being and commitment, and mission success are dependent upon the application of CQ at the national level; that is the ability of the CF and its leaders to understand and meet the expectations of Canadian society in terms of its treatment of its own members as well as its performance with external actors, organizations, groups, and nations. The military ethos guides the moral application of CQ across all dimensions of the effectiveness model, as well as the national, international, host-nation, and enemy domains.[21] As an essential aspect of CF professional and leadership capacity, CQ permeates the effectiveness model in contributing to mission success.

UNDERSTANDING CULTURE

CQ becomes particularly essential to mission success when the CF operates within international cultures and societies, and across both domestic and internationally-based networks and organizations. Numerous models and tools are available to provide preliminary insight into the cultures of organizations and societies around the world. Values-based surveys are perhaps the most well-known, in part because they produce large amounts of information. While the data available from sources such as these surveys can be extremely valuable, it is important to treat all data as one of several potential sources of information rather than the solution to understanding culture.[22] For example, within the comparative frameworks of international values-based surveys, the nation state is used as a proxy for culture thus oversimplifying and misrepresenting cultural boundaries, and culture is evaluated from an external position using indicators developed by cultural outsiders.

In a critique of the work of Geert Hofstede, for example, Rachel Baskerville notes that the *Encyclopedia of World Cultures* identifies 147 Native American cultures and nine North American folk cultures even though all of these cultures are subsumed under the internationally recognized countries within which they are situated. Alternatively, sociologists and anthropologists are more likely to rely upon ethnographic studies using fieldwork methods such as participant observation and interpretive/non-quantitative approaches to gain understanding within localized cultural settings.

Notwithstanding the criticism of values-based surveys, several of these tools have generated data that provides a basic understanding of some cultural differences. Hofstede's work, covering over 100,000 employees of a large multinational corporation spread across fifty countries has been, perhaps, the most widely used to understand differences across culture.[23] It suggests five value dimensions representing cultural variation,[24]

- ***individualism-collectivism:*** the strength of loyalty to the groups/organizations with which members identify;
- ***power distance***: the extent to which large differentials of power, for example, between a boss and a subordinate or between a higher-status and lower-status person, are expected and tolerated;
- ***uncertainty avoidance***: the extent to which the culture emphasizes focusing on ways to reduce uncertainty and create stability, for example, having clear written rules and procedures or strong norms to guide actions;

- ***masculinity/femininity***: the balance between the traditional "male" goals of ambition and achievement and "female" orientations to nurturance and interpersonal harmony, for example, the balance between seeking promotion at work and having good relationships with others; and

- ***long-term orientation***: the time perspective in a society for the gratification of needs. Long-term orientation implies an emphasis on virtuous living in the world, and short-term orientation places emphasis on finding the truth.[25]

In discussing Hofstede's framework within the context of CQ, David Thomas and Kerr Inkson indicate that the individualism-collectivism dimension is the most useful and powerful dimension of culture that has been identified,[26] and thus warrants a bit more explanation in this discussion. The individualism-collectivism dimension refers to:

> ...the relationship between the individual and the collectivity which prevails in a given society. It is reflected in the way people live together – for example, in nuclear families, extended families, or tribes; and it has all kinds of value implications. In some cultures, individualism is seen as a blessing and a source of well-being; in others, it is seen as alienating.[27]

In other words, this value dimension emphasizes the degree to which individuals express and prize independence and standing out or, alternately, the degree to which they are focused on fitting in with the group. For example, in Canada there is typically an expectation that children move out of their parents' home once they reach adulthood; that they need to "get out in the world" and "stand on their own two feet". This tends to occur around the ages of 18-24. However, the cultural norm in many parts of South America is for children to continue to live with their parents until well into their late twenties and early thirties. Similarly, in some cultures, the extended family or the community assume much greater importance than the nuclear family.

Interestingly, Hofstede connects this particular value dimension with identity. Referring to individualism, he notes that "[t]he central element in our mental programming involved in this case is our self-concept."[28] In other words, Hofstede makes a more direct connection between an individual's identity and the culture in which that identity was formed. Generally speaking, this value dimension implies that those in individualistic cultures will tend to form identities that are inwardly focused; that is, their values, beliefs and practices will take on a much more singular or personalized interpretation of their cultural or group values. Conversely, those in collectivistic cultures will tend to form identities that are outwardly focused; that is, their personal values and practices will be more synchronized with those of the social group and the environment in which they live. Recognizing the way in which a person's self-concept is formed is a crucial component of cross-cultural understanding or CQ.

Another well-known culture tool, the World Values Survey (WVS), is one of the most comprehensive tools in terms of the number of societies involved and the information that has been collected over time. The WVS has been used to gather data since 1981 from 85 nations representing 85 per cent of the world's population. This data has been used to track and compare human development on socio-economic, cultural, and institutional dimensions. WVS experts Ronald Inglehart and Christian Welzel note, "the basic values and beliefs of the publics of advanced societies differ dramatically from those found in less developed societies," and that these values are changing as socio-economic development takes place.[29] From the perspective of the WVS, population, human capital, workforce, social status, living conditions, science and religion, mass media, government, social protection, family structures, sex roles, and cultural values each provide clues to societal change.[30]

In a security environment, it is not sufficient to understand only one aspect of the culture, or to interpret culture without taking the impact of other influencing and integrated cultures into account. The available information must be used to develop further understanding and insight into tactical, operational, and strategic operations. Just as the CF reflects Canadian values and assumptions but remains unique from Canadian society in many ways, other Canadian-based communities, networks, and organizations, will reflect general Canadian cultural orientations and at the same time differ in some of the ways that they think, plan, and act. Host cultures interact with many expressions of other cultures that operate to realize an endless array of goals, such as: realize financial profit; provide food and shelter for families; prevent and treat disease; establish economic infrastructure; promulgate political messages; recruit and train soldiers; establish social equality; and inflict harm. These relationships result in complex systems of interactions with intended and unintended consequences and opportunities.

As discussed in Chapter 1, complex systems are omnipresent within cultural milieus. It is imperative that the CF have the capacity to operate as an adaptive system; that is, a system that is capable of learning and evolving to adapt to influences of varying systems and contexts. Importantly, the boundaries of a given system are established only by the interpreter, thus demanding high levels of CQ to conduct continuous analysis and interpretation depending upon the particular challenge and context. The CF and its various entities represent, to various degrees, a complex system of systems that is adaptive and open in some ways, and less complex and bound by inflexible rules in others. Table 2-1, summarizes and reinforces the features of complex systems.

1. Complex systems consist of a large number of elements.
2. These elements interact dynamically.
3. Interactions are rich; any element in the system can influence or be influenced by any other.
4. Interactions are non-linear.
5. Interactions are typically short-range.
6. There are positive and negative feedback loops of interactions.
7. Complex systems are open systems.
8. Complex systems operate under conditions far from equilibrium.
9. Complex systems have histories.
10. Individual elements are typically ignorant of the behaviour of the whole system in which they are embedded.

TABLE 2-1: FEATURES OF COMPLEX SYSTEMS[31]

ORGANIZATIONS AND NETWORKS

As indicated earlier, variations of culture exist within and across societies within national, community, and organizational/institutional domains. Culture is often influenced by, and reflective of, identifiable political, economic, and social structures; however, culture also flourishes within and permeates networks that do not have identifiable hierarchies and structures. From this perspective, a complex-system model is a particularly valuable lens through which to frame understanding of cultural institutions, organizations, and networks.

In addition, there are various tools and frameworks available to further identify and define a range of cultural characteristics. The work of David S. Alberts and Richard E. Hayes is particularly relevant to security-focused cultures, such as the military, in the information age. Alberts and Hayes compare hierarchies to edge organizations on the basis of command, leadership, control, decision-making, information, predominant information flows, information management, sources of information, organizational processes, and individuals at the edge,[32] as presented in Table 2-2.

These characteristics of military organizations and security networks can be particularly important to the effectiveness of operations which are predicated on a thorough understanding of an adversary's command and control, and decision-making processes. Edge organizations, according to Alberts and Hayes, are those in which command is not coupled with control as is the case in traditional military organizations. Command establishes the initial conditions and provides overall intent; however, control is an emergent property in response to initial conditions, the environment, and adversaries. Power is decentralized and those who are responsible for dealing with situations and accomplishing tasks have the means and opportunities to respond, thus increasing agility when faced with uncertain and unfamiliar situations.[33]

	Hierarchies	Edge Organizations
Command	By directive	Establishing conditions
Leadership	By position	By competence
Control	By direction	An emergent property
Decision-making	Line function	Everyone's job
Information	Hoarded	Shared
Predominant Information Flows	Vertical, coupled with chain of command	Horizontal, independent of chain of command
Information Management	Push	Post – Pull
Sources of Information	Stovepipe monopolies	Eclectic, adaptable marketplaces
Organizational Processes	Prescribed sequential	Dynamic concurrent
Individuals at the Edge	Constrained	Empowered

TABLE 2-2: CHARACTERISTICS OF HIERARCHIES AND EDGE ORGANIZATIONS

Institutions and cultural networks can also be understood by examining the institutional work that creates, maintains, and in some cases, disrupts the rules and logic practiced by the institution and its professionals. In an extensive review of empirical organizational research, Thomas B. Lawrence and Roy Suddaby[34] describe the processes and activities that create, maintain and disrupt institutions through the work of its members. For example, they describe the importance of constructing rule systems that confer status or identity, define boundaries of membership or create status hierarchies.[35] Processes which disrupt institutions, they postulate, include disconnecting sanctions and rewards, undermining moral foundations, and undermining the core assumptions and beliefs of the organization.[36]

Frameworks can be applied to many visual and observable practices within organizations as a first step to understanding cultural cues in unfamiliar domains. For example, in most cases, it would be relatively easy to identify the rule system that guides behaviour in a given organization if you have access to the language and written rules. However, discerning the relationship between those rules, and the cultural interpretations that result in practice and behaviour becomes increasingly complex without ready access and given the overlapping practices and interests at play. Given appropriate time, resources, and expertise, qualitative approaches to cultural analysis such as ethnographic studies can also generate valuable CQ-related information. CQ provides the basis for continuous interpretation and adaptation to knowledge and experience to facilitate effective negotiation of cultural challenges.

SUMMARY

The development of effective CQ among individual members and across the CF is a key enabler of adaptability and mission success. Importantly, CQ is distinct from cultural awareness and cannot be entirely achieved through culture-specific knowledge. While cultural awareness contributes to CQ, awareness is not sufficient in and of itself to address challenges in an unfamiliar milieu. In addition to culture-specific knowledge, frameworks for understanding the culture of a society, institution or network are important contributors to cultural knowledge. Importantly, as indicated in Chapter 1 and emphasized by this discussion, culture, its patterns, attitudes, beliefs and motivations, is not unique to nations or societies, but in reality exists across and within groups, teams, and networks, whether virtual and dispersed or local and structured. Integrating such knowledge with an ability to untangle and interpret the essential elements of cultural meaning and processes is an essential contributor to CQ. Chapter 3 pursues the relationship between national strategy and policy, CF operations, and CQ.

NOTES

1 Emily Spencer, *Crucible of Success: Applying the Four CQ Domain Paradigm, The Canadian Forces in Afghanistan as a Case Study*, (Kingston, ON: Canadian Forces Leadership Institute, 2007), 3. This definition of CQ was expanded by a Canadian Forces Leadership Institute CQ project team to include "or a range of activities."

2 Rachel F. Baskerville, "Hofstede never studied culture," *Accounting, Organizations and Society*, 28 (2003), 1-14.

3 Ronald Wright, *A Short History of Progress*, (Toronto, ON: House of Anansi Press Inc., 2004), 32.

4 Ibid.

5 Robert Kegan, *The Evolving Self: Problems and Process in Human Development*, (Cambridge: Harvard University Press, 1982).

6 Agency, in the sociological sense, refers to the power of individuals to operate independently of the constraints of social structures, thus conveying the volitional, purposive nature of human activity as opposed to its determined, constrained aspects.

7 P. Christopher Earley and Soon Ang, *Cultural Intelligence: Individual Interactions Across Cultures*, (Stanford, California: Stanford University Press, 2003) 9-10. In addition, the Center for Leadership and Cultural Intelligence, and the Cultural Intelligence Center in Singapore present CQ as the integration of strategy, knowledge, motivation, and behaviour. See Center for Leadership and Cultural Intelligence, <http://www.cci.ntu.edu.sg/index.html>, (2004); and Cultural Intelligence Center, <http://www.culturalq.com/index.html>, (2007); James P. Johnson, Tomasz Lenartowicz, and Salvador Apud define CQ as the integration of attitude,

skills, and knowledge. See James P. Johnson, Tomasz Lenartowicz, and Salvador Apud, "Cross-cultural competence in international business: toward a definition and a model," *Journal of International Business Studies* 37(4, 2006), 534; and David C. Thomas and Kerr Inkson define CQ as the combined effect of knowledge, skills, and mindfulness. See David C. Thomas and Kerr Inkson, *Cultural Intelligence: People Skills for Global Business*, (San Franciso: Berrett-Koehler Publishers, Inc., 2005); and David C. Thomas, "Domain and Development of Cultural intelligence: the Importance of Mindfulness," *Group and Organisation Management* 31 (2006).

8 Thomas, "Domain and Development of Cultural intelligence," 85-86.

9 It is important to note that this illustration does not represent a model of CQ in the sense that it has not been validated by social scientific measurement. For an overview of the measurement and validation of CQ models see Kimberly-Anne Ford and Karen D. Davis, *Cultural Intelligence, Emotional Intelligence and Canadian Forces Leader Development: Concepts, Relationships and Measures*, (Kingston, ON: Canadian Forces Leadership Institute and Defence Research and Development Canada, 2007), CFLI Technical Memorandum 2007-01.

10 John P. Coles, "Incorporating Cultural Intelligence into Joint Doctrine," *Iosphere: Joint Information Operations Center* (Spring 2006), 7.

11 Thomas, "Domain and Development of Cultural Intelligence," 83.

12 Earley and Ang, *Cultural Intelligence*, 10.

13 Ibid., 11.

14 Ford and Davis, *Cultural Intelligence, Emotional Intelligence and Canadian Forces Leader Development*.

15 Thomas, "Domain and Development of Cultural Intelligence," 78-99.

16 Ibid, 85.

17 The CF Professional Development Framework was developed to reflect the leader capacities and attributes required to operationalize CF leadership doctrine. See Robert W. Walker, *The Professional Development Framework: Generating Effectiveness in Canadian Forces Leadership*, (Kingston, ON: Canadian Forces Leadership Institute, 2006), Technical Report 2006-01.

18 For a comparison of competencies and attributes that comprise the CF PDF, CQ, and emotional intelligence, see Ford and Davis, *Cultural Intelligence, Emotional Intelligence and Canadian Forces Leader Development*.

19 These characteristics adapted from assumptions underlying CQ outlined in Chay Hoon Lee and Klaus J. Templer, "Cultural Intelligence Assessment and Measurement," *Cultural Intelligence: Individual Interactions Across Cultures*. P. C. Earley and S. Ang (eds). (Stanford, California: Stanford University Press, 2003), 185-208.

20 *Leadership in the Canadian Forces: Conceptual Foundations (2005), Leadership in the Canadian Forces: Doctrine* (2005), *Leadership in the Canadian Forces: Leading the Institution* (2007), and *Leadership in the Canadian Forces: Leading People* (2007), (Kingston: ON, Canadian Forces Leadership Institute).

21 These domains are presented here as a general description of a potential range within which CQ is applied. As discussed in Chapter 4 of this volume, it is also important to consider organizational and institutional cultures when applying CQ. For discussion of these four domains as a potential paradigm, see Emily Spencer, *Crucible of Success*.

22 For a discussion of the shortcomings of Hofstede's value-based survey approach to studying culture, see Rachel F. Baskerville, "Hofstede never studied culture," *Accounting, Organizations and Society*. 28 (2003), 1-14.

23 Thomas and Inkson, *Cultural Intelligence*, 31.

24 Ibid., 34.

25 Geert Hofstede. "Images of Europe," *The Netherlands Journal of Social Sciences*, 30 (1994) 63-82.

26 Thomas and Inkson, *Cultural Intelligence*, 36.

27 Geert Hofstede, *Culture's Consequences: International Differences in Work-Related Values*, (Beverly Hills: Sage Publications, 1980), 213.

28 Ibid., 215.

29 Ronald Inglehart and Christian Welzel, *Modernization, Cultural Change, and Democracy: The Human Development Sequence*, (Cambridge University Press, 2005), 1.

30 Ronald Inglehart and Pippa Norris, *Rising Tide: Gender Equality and Cultural Change Around the World*, (Cambridge University Press, 2003), 12-14.

31 P. Cilliers, *Complexity and Postmodernism: Understanding Complex Systems*, (London: Routledge, 1998) cited in Steve Maguire, Bill McKelvey, Laurent Mirabeau and Nail Öztas, "Complexity Science and Organizational Studies," *The Sage Handbook of Organization Studies* (2nd Edition). Stewart R. Clegg, Cynthia Hardy, Thomas B. Lawrence and Walter R. Nord (eds.), (London: Sage Publications, 2006), 165-214.

32 David S. Alberts and Richard E. Hayes, *Power to the Edge: Command…Control…in the Information Age*. (DoD Command and Control Research Program, 2003), 218.

33 Ibid., 216-219.

34 Thomas B. Lawrence and Roy Suddaby, "Institutions and Institutional Work," *The Sage Handbook of Organization Studies* (2nd Edition). Stewart R. Clegg, Cynthia Hardy, Thomas B. Lawrence and Walter R. Nord (eds.), (London: Sage Publications, 2006). 215-254.

35 Ibid., 223-224.

36 Ibid., 237-238.

RECOMMENDED READING

Alberts, David S. and Richard E. Hayes, *Power to the Edge: Command...Control...in the Information Age*. United States: Department of Defense, Command and Control Research Program, 2003.

Coles, John P. "Incorporating Cultural Intelligence into Joint Doctrine," *Iosphere: Joint Information Operations Center* (Spring 2006).

Earley, P. Christopher and Soon Ang. *Cultural Intelligence: Individual Interactions Across Cultures*. (Stanford, California: Stanford University Press, 2003).

Ford, Kimberly-Anne and Karen D. Davis. *Cultural Intelligence, Emotional Intelligence and Canadian Forces Leader Development: Concepts, Relationships and Measures*. Kingston, ON: Canadian Forces Leadership Institute and Defence Research and Development Canada, 2007. CFLI Technical Memorandum 2007-01.

Lawrence, Thomas B. and Roy Suddaby. "Institutions and Institutional Work," *The Sage Handbook of Organization Studies* (2nd Edition). Stewart R. Clegg, Cynthia Hardy, Thomas B. Lawrence and Walter R. Nord (eds.), London: Sage Publications, 2006.

Maguire, Steve, Bill McKelvey, Laurent Mirabeau and Nail Öztas. "Complexity Science and Organizational Studies," *The Sage Handbook of Organization Studies* (2nd Edition). Stewart R. Clegg, Cynthia Hardy, Thomas B. Lawrence and Walter R. Nord (eds.), London: Sage Publications, 2006.

Thomas, David C. and Kerr Inkson. *Cultural Intelligence: People Skills for Global Business*. San Franciso: Berrett-Koehler Publishers, Inc., 2005.

Thomas, David C. "Domain and Development of Cultural Intelligence: the Importance of Mindfulness," *Group and Organisation Management* 31 (2006).

CHAPTER 3

CULTURAL INTELLIGENCE AND STRATEGIC CULTURE

Dr. Bill Bentley

Policy-makers, strategists and, in fact, all those involved in implementing Canada's national security policy need to be thoroughly familiar with the national strategic culture within which they operate. Canada normally operates within alliances or coalitions, and although the strategic cultures of our allies or partners are often quite similar there are still differences, often significant ones. Furthermore, understanding something about the strategic culture of our adversary is critical to mission success.

This chapter will define and discuss the concept of strategic culture and outline the factors that contribute to its creation and maintenance. With these factors in mind, the chapter will go on to describe, briefly, Canada's strategic culture to illustrate the kind of analysis that would need to be conducted in order to come to grips with another country's or political community's strategic culture. Strategic culture is only one element of CQ, but it is an important one.

STRATEGIC CULTURE

The noted strategic theorist, Colin S. Gray, has recently written that war is a relationship between belligerents, not necessarily states. Warfare is the conduct of war, primarily, though not exclusively, by military means. The two concepts are not synonymous. There is more to war than warfare.[1] This reflects Gray's deep understanding that, as Carl von Clausewitz demonstrated in his magisterial work *On War*, "war is merely the continuation of a nation's policy with the admixture of other means."[2]

War is, therefore, the all-inclusive concept led by politics and the policy level. Warfare is conducted at the operational and tactical levels. Strategy, also a key element of warfare, is the bridge between war and warfare. It is the art of distributing and applying military means, or the threat of such action, to fulfill the ends of policy. Strategy is dynamic, iterative and non-linear. War should be viewed as a system; the sub-systems of strategy, operational art and tactics are nested within the guiding system-policy.

Clearly then, to speak of a nation's strategic culture goes beyond considering only the military component of national security policy, strategy, operational art and tactics, as important as these components are. In fact, in most states, the country's strategic culture is in part defined by the very nature of its civil-military relations and the tensions inherent in this aspect of national security; that is, the interface between policy and strategy, war and warfare. Therefore, in considering any nation's strategic culture it is necessary to integrate the concepts of war and warfare: the way its inhabitants, as a nation, have understood war and conflict philosophically, culturally and politically and how the nation, mainly through its military instrument, has conducted warfare.

Strategic culture refers to the socially transmitted habits of mind, tradition and preferred methods of operations that are more or less specific to a particular geographically-based security community. It is a product of a particular national historical experience which has been shaped by a more or less unique, though not necessarily unvarying geographic context. Each strategic culture is inclined to erect what purports to be general theories on the basis of national historical experience and circumstances.

With this general understanding, the chapter defines strategic culture as follows:

> *An integrated system of symbols (argumentation structure, languages, analogies and metaphors, etc.) that acts to establish pervasive and long-term strategic preferences by formulating concepts of the role and efficacy of military force in political affairs. The strategic culture thus established reflects national preconceptions and historical experience as much as it does purely objective responses to any given threat environment.*[3]

Strategic culture as a "system of symbols" consists of basic assumptions[4] about the orderliness of the strategic environment; in other words, about the role of war in human affairs, whether it is inevitable or an aberration; about the nature of the adversary and the threat it poses – zero-sum or variable-sum; and about the efficacy of the use of force; that is, the ability to control outcomes and to eliminate threats and the conditions under which applied force is useful.

In his book, *Battle: A History of Combat and Culture*,[5] John A. Lynn talks about how a nation possesses a "discourse" on war. This discourse is culturally determined and is one helpful way of viewing the concept of strategic culture. It is a long-term, slow growth phenomenon not particularly dependent on specific individuals or even any single, significant event. When two or more nations or communities engage in armed conflict we might say that a meta-discourse is being played out in a confrontation of all of the assumptions mentioned in the previous paragraph. Importantly, CQ includes both the empathetic ability to discern the nature of the adversary's assumptions and the ability to analyze them in terms of the factors of strategic culture.

Factors of Strategic Culture

In their book, *The Making of Strategy*, Williamson Murray, Alvin Bernstein, and MacGregor Knox argue that there are four major factors that contribute to the evolution of a strategic culture:[6]

- ***Geography*** – The size and location of a nation are crucial determinants of the way policy-makers and strategists think about security and strategy. This, of course, includes all of the national resources available to the nation.

- ***History*** – Historical experience influences strategic culture almost as strongly as geography.

- ***Religion, Ideology and Culture*** – Taken together, these three terms comprise something the Germans have captured in a single expressive word – *Weltenschung* (world view or outlook on the world) the influence of this concept on strategic culture is both elemental and vast, according to Murray *et al*.

- ***Governance*** – The structure of government and military institutions plays a crucial role in the development of strategic culture.

Colin S. Gray's own list, outlined in his *Modern Strategy*, is remarkably similar:[7]

- Geography, the most fundamental of the factors that condition national outlooks on security problems and strategic solutions.

- History.

- The influence of different national cultures upon choices for, and performance in, statecraft, war and warfare.

- Relative technological competence is important. However, technicism refers to the disorder when that which is only technical displaces, and effectively substitutes for, that which has to be considered tactically, operationally and strategically in far more inclusive analysis.

In each of these models the factors of geography, history, governance and technology largely fit into the knowledge dimension CQ described in Chapter 2. However, when considering Murray et al's, "Religion, Ideology and Culture," or Gray's, "different national cultures," the entire CQ model must be employed. These strategic culture factors also demand the cognitive, motivational and behavioural dimensions that relate to the analyst's ability to empathize, understand and communicate with the ideology and culture in question. As discussed in Chapter 1, it is here that a holistic systems approach to thinking about the issue is indispensable. Therefore, as with any other cultural entity, strategic culture must be approached using systems thinking and the models and concepts developed in Chapter 2.

THE GENERAL SYSTEM OF WAR AND CONFLICT

To consider Canada's strategic culture, five factors, at a minimum, should be analyzed: geography, history, culture, governance and technology. However, before dealing with each of these factors in turn as they apply to Canada, there is an additional requirement to understand the generic structure of security problems and military operations that are addressed through the unique prism of a nation's strategic culture.

As illustrated in Figure 3-1, wars and conflicts in general are conducted at four levels – political, strategic, operational (theatre) and tactical, with each level nested or sitting within the context of the other in descending order from the political. It is this latter level that gives context to all the activities of all levels aimed at the same objectives and enables coherence between them.

FIGURE 3-1: THE GENERAL SYSTEM OF WAR AND CONFLICT

Strategy is an expression of the aim and its links to the overall purpose and context of the conflict, together with the limitations on action, the circumstances of which flow from the political purpose. It will describe the desired pattern of events together with the measures intended to achieve this pattern and will dictate the allocation forces and resources. Analytically, there are essentially two strategic systems which relate conceptually to the political purpose of the war or conflict. If the goal is very ambitious and requires the unconditional surrender of the opponent, the strategy is one of decisive military victory or of annihilation. If, on the other hand, the goal is limited, the strategy is bipolar: one pole being battle and the other, co-equal pole, being the simultaneous or sequential use of other non-military means to bring about a mutually satisfactory settlement. World War I (WWI) and World War II (WWII) are examples of the first strategic system, while Korea would be a good example of the second. The suitability and utility of either type of strategy are very much the product of the strategic culture of the nation in question.

In Western military thought from the Napoleonic era through the 1990s, the strategy of annihilation was the foundational doctrine. To be sure, the British, the French, the Germans and the United States (US) fought limited campaigns and were extensively involved in policing their respective empires. However, every staff/war college curriculum focused almost exclusively on how to fight major wars aimed at decisive victory. It was, in fact, a pervasive and powerful element in the strategic culture in all these nations. Only with the advent of nuclear weapons did the defence community of the Western World begin to fully explore the validity and implications of the alternate strategic system – the bipolar strategy. It was then that strategic culture began to change.

The theatre or operational level of war or conflict is conducted in the theatre of operations – a geographical area containing in its military and political totality an objective that on achievement alters the strategic situation to advantage. The theatre commander must make a plan, a campaign, which designates the path towards the final end state and orchestrate the activities of the whole command to achieve tactical objectives, thus taking the command, as a whole, in the designated direction.

Conceptually, at the operational level, warfare is conducted through either attrition or manoeuvre; a preference or necessity shaped to one degree or another by all of the factors of strategic culture enumerated above. Examples here might be the French in the interwar years with a doctrine of attrition that stressed firepower at the outset of a campaign, switching to manoeuvre only after the enemy had been thoroughly blooded. Conversely, the Germans devised a doctrine that stressed operational manoeuvre, preferring mass and mobility in the right proportions over mass firepower. For all operational-level commanders, the challenge is to get the right balance between mass and mobility – the definition of operational manoeuvre.

Engagements (military and non-military), battles and fights are at the tactical level. The essence of all tactics is fire and movement. This is tactical manoeuvre, which in essence boils down to attrition by destruction. The basic tactical dilemma here is to find the correct balance between how much effort to apply to striking the opponent to achieve the mission and how much to countering his blows and undermining his ability to sustain himself.

CANADA'S STRATEGIC CULTURE

Canada's strategic culture addresses each of the levels of war in an interactive way, and in a manner unique to the country. Five key factors, considered together, account for strategic culture at each level.

Strategic Culture: Geography

Canada's geography has always posed a particular dilemma. Initially, of course, membership in the British Empire and Britain's command of the sea reassured Canadians in security terms. By the turn of the 20th century, however, direct British support could no longer be counted on. At the same time, for as long as reasonably good relations could be maintained with the US, geography seemed an almost insurmountable barrier to any direct military threat. At the same time, sharing a continent and a common border with such an economically and culturally powerful neighbour also posed a significant non-military threat in terms of national sovereignty and in extreme scenarios, actual absorption. Successfully maintaining good relations with the US also meant that Canada could rely on the militia paradigm throughout much of Confederation's history. This accorded well with historical experience and political preference (governance) up until WWII. WWI was a partial exception and by 1917, the Canadian Army in Europe had to been acknowledged as a battle-hardened, professional force. Nonetheless, the return to a very small standing military force, ostensibly backed up by a large militia, demonstrated Canada's political/strategic preference very clearly.

Technology itself further complicated the matter with the advent of nuclear weapons and ballistic missiles. Now Canada was directly implicated in US security and a delicate balance was necessary between reassuring its neighbour and limiting the Canadian defence effort as much as possible under the circumstances. Thus, aerospace defence, and especially participation in the North American Aerospace Defense Command (NORAD), became priorities for Canadian defence policy.

Geography also conspired with history and culture to lead Canada to seek multilateral solutions to security/strategic problems in order to limit the influence of American preferences on Canadian policies. Additionally, multilateralism created "niche" spaces for middle powers like Canada to make viable contributions to international security. Over the years, Canada rather successfully maintained the required balance. Geography alone meant that a minimum defence effort had to be made, but ultimately such an effort gravitated toward something nearer to a diplomatic payment as opposed to a full-scale military response. With the demise of international communism, the potential for the influence of geography on Canada/US relations, and therefore Canada's strategic culture, to lessen was certainly there. However, the advent of global terrorism has reaffirmed the importance of geography in the shaping of strategic culture.

Strategic Culture: History

Historical contingency in the pre-Confederation era has contributed significantly to Canada's strategic culture. Under both French and English rule, the militia paradigm was entrenched in Canada. Furthermore, Canadian militiamen logically preferred to take the fight to the enemy rather than fight at home. Here, in embryonic form, was the Canadian preference for expeditionary forces, usually small and only when required, rather than maintaining large standing forces at home.

After Confederation, the tension between involvement in European affairs, mainly British, and the progress towards autonomy and finally, full sovereignty was intense. There was a large and influential constituency in Canada, political, military and social from before WWI and extending well into the 1930s, which called for close and large-scale support for the "Mother Country." This was largely provided for in WWI through the provision of the First Canadian Corps which established the reputation of Canadian soldiers for tactical excellence and fighting skills. However, after 1918 the political imperative for sovereignty together with the cost of developing the vast Canadian country meant that governments sought to limit military engagement with the Empire, reduce military spending to the lowest limit and to restrict military-to-military co-operation between the Canadians and the British to the maximum extent possible. This in itself was somewhat paradoxical since it was Canadian military prowess and the enormous sacrifices made in WWI that helped cement Canadian nationhood.

Growing political autonomy during this period brought with it a sense of international responsibility. Membership in the League of Nations was one response, although active involvement here was counterbalanced by a geographically induced sense of isolation and a continuing political preference for miniscule standing military forces and a large, but quiescent militia force. What military activity occurred was largely restricted to aid to the civil power in the 1920s depression-era, support for government programs and a great deal of military support for nation building inside Canada itself.

Significantly however, ever-closer ties were developed on a military-to-military basis between the Canadian army and navy and their British counterparts. Thus, while the government sought to limit involvement in European affairs, military outlook and doctrine were heavily influenced by British strategic concerns and doctrine. This would be a recurring theme in Canadian strategic culture with both the US and the ongoing debate of the primacy of either the United Nations (UN) or the North Atlantic Treaty Organisation (NATO) in the post-WWII era.

WWII was a dramatic and decisive turning point for Canada, politically, socially, economically and militarily. Over one million Canadians out of a population of 11 million served, and served honourably. Canada's reputation for tactical excellence was again consistently demonstrated. By the end of the war, Canada was ready for full-scale international commitment first through the UN and then in 1949 by membership in NATO. Multilateralism became an entrenched cornerstone of Canadian foreign and defence policy with the concomitant requirement for relatively large military forces forward deployed and employed in Korea, Europe, and a growing number of UN missions around the globe.

The threat of nuclear war between the Soviet Union and the US was the preeminent factor driving Canada's foreign and defence policy after WWII. Every effort was made to contribute to global stability unilaterally, as in Canada's peacekeeping involvement in Vietnam or, in a much larger way, through the UN. Although humanitarian motives for extensive involvement in UN operations cannot be completely discounted, in fact, Canadian actions were primarily to help rebuild US-United Kingdom (UK) relations (for example, the United Nations Emergency Force (UNEF) in the Middle East in 1956 or through participation in the United Nations Force in Cyprus (UNFICYP) to shore up NATO's southern flank). Stabilizing hot spots throughout the Cold War to minimize potential confrontations between NATO and the Warsaw Pact largely accounted for Canada's continued involvement in the Middle East, Africa and Kashmir.

It was, however, Canada's NATO involvement on the ground that shaped the Canadian military doctrine during the 40-year period between the early 1950s and the early 1990s. Close ties with first the British in northern Germany and then the Americans in central Germany meant that Canadian military officers engaged heavily in foreign strategic and operational theory and doctrine. In fact, a key element in "Canada's way of war" was most clearly manifested here. In addition to its multinational nature, Canada's commitment was always restricted to the tactical level. Retrospectively, this had, of course always been the case since the Boer War, through WWI, WWII and Korea. This fact has two profound implications for Canadian strategic culture. First, little concerted thought was ever given to the operational level, operational art or applied military strategy or doctrine. At the same time, tactical elements were assigned to non-Canadian operational command in service packages – Navy to Navy, Army to Army and Air Force to Air Force. This led to a very distinct "strong single service" ideology and its consequence would be felt significantly after the Cold War.

The heavy reliance on US, British and NATO doctrine at the operational and strategic levels meant that the Canadian military always viewed their tactical contributions as part of a strategy of annihilation. Despite extensive peacekeeping experience, curricula at the Canadian staff colleges and the National Defence

College were dominated by doctrines that foresaw large-scale, high-intensity warfare in which NATO sought decisive victory over the Warsaw Pact. At the theoretical and doctrinal levels, there seemed little need to study bipolar strategy.

With the demise of the Soviet Union and the consequent paradigm shift in the character of conflict and war in the 1990s, the reliance on a strategy of annihilation was slowly eroded in the US, UK and NATO. It would take another 10 years for this process to have an appreciable impact in Canada and the Canadian Forces.

Strategic Culture: Religion, Ideology and Culture

Religion, ideology and culture have had a strong influence on Canadian strategic culture in two important ways. As a bilingual, liberal constitutional democracy, Canada always retained strong political ties with the UK and the US, and indeed most western countries. Sharing a common ideology meant that Canada found any kind of totalitarianism anathema to the desired international order. Thus, it was a foregone conclusion that fascism and Nazism would be resisted, although at what level remained a delicate political issue. Similarly, communism was to be contained and ultimately defeated. In the 20th century this led to the raising and maintaining of large, standing professional forces in WWII and throughout the Cold War. In keeping with Canadian strategic preference and as an integral part of Canada's strategic culture, these forces were forward deployed in Germany. The strength of this factor, in practical terms, waxed and waned with the perception of the power of the competing ideology. Thus, during periods of *détente*, defence spending declined. Once the threat disappeared, defence spending plummeted. It remains to be seen whether the ideologies of ethnic extremism, Islamic fundamentalism and terrorism will continue to elicit a similar proactive response.

However, by far the most important contribution this factor has made to strategic culture is the bicultural nature of Canadian confederation. It has been argued that peacefully managing English-French relations throughout Canada's history predisposes Canadian political culture to tolerance, mediation and patience. Consequently, so the argument goes, Canadians are "helpful fixers" on the international scene. In short, we make good peacekeepers! While not definitive, this line of thought is plausible. However, biculturalism helped to shape "Canada's way of war" in a much more direct way. As national unity has been paramount to Canadian political leaders since Confederation, the threat of national disintegration has been an overriding concern. French-Canadian reluctance to wholeheartedly support the British Empire in any significant military way from the Boer War onwards meant that military commitments had to be managed very carefully. The conscription crises of WWI and WWII were salutary warnings to Canadian policy-makers. A largely Roman Catholic French Canadian population meant that support for anti-communism generally remained strong throughout the Cold War. The rise of separatism in the 1980s and 1990s complicated the matter further and Canada's military involvement on any large scale anywhere remains subject to at least a careful survey of national attitudes and the English-French split on these matters. Certainly it appears that strong Francophone objections to Canada's participation in the US-led invasion of Iraq were a significant factor in federal decision-making.

Religion, ideology and culture in Canada also coalesced around the issue of nuclear weapons. Canadians were very ambivalent concerning their utility and generally rejected the idea that Canada should ever become a nuclear weapons state. Although Canada did, in fact, possess such weapons in the 1950s and 1960s, they remained a burning political issue and were ultimately rejected.

The growing multicultural complexion of Canada since the 1960s has also helped alter and shape its strategic culture. Whereas previously the Canadian strategic focus has always been on Europe, politicians have increasingly had to take into account Canadian attitudes towards instability and conflict around the globe. Despite the slow-growing nature of strategic culture, this factor will invariably play a significant role in how Canada views security in the future.

Strategic Culture: Governance

There has always been a certain tension in Canadian civil-military relations; indeed, this is a characteristic of all western democracies. In Canada, these tensions have resulted in differing views on the utility of force in international politics and above all in the political desire to maintain a tight control over military policy and keep defence spending at a minimum. If civilian policy-makers and officials seemed indifferent or unresponsive to perceived threats, military officers often appeared to civilians to be overly zealous in their demands for action. All would do well, however, to remember that in a liberal democracy, the military can propose the level of armaments necessary to have a certain probability of successful defence from one's enemies, but only the civilian can say for what probability of success society is willing to pay. The military can describe in some detail the nature of the threat posed by a particular enemy, but only the civilian can decide whether to feel threatened and, if so, how, or even whether, to respond. The military assesses the threat, the civilian judges it.

In Canada's strategic culture, these issues are addressed sporadically through the issuance of Defence White Papers (1947, 1964, 1970, 1987, 1994, 2004 and 2005) that set the context for political and bureaucratic processes to manage civil-military relations in the intervening periods. The establishment of the single Department of National Defence in the immediate aftermath of WWI was in large part a response to these considerations. Similarly, after the massive effort in WWII, measures were taken in the postwar years to re-establish tight political oversight and control spending. Unification and integration were major governance decisions intended to shape the forces for rapid, expeditionary-type operations and streamline the structure internally, as well as reduce access to the Minister to one military officer, namely the new Chief of the Defence Staff (CDS).

These measures, however, actually did very little to reduce the "strong single service" ideology and at the same time perpetuated the practice of operating only at the tactical level. At the end of the Cold War and throughout the 1990s, it became increasingly apparent that neither characteristic of the prevailing strategic culture was well suited to the new challenges of global disorder and terrorism. The Defence White Paper of 2005 attempted to plot a new political-military course and the CDS transformation project was designed to implement this in detail. The first principle of transformation emphasized movement from service cultures to a CF culture. Structurally, for the first time in Canadian history, truly joint, operational-level headquarters were established and an explicit strategic staff organization was put in place. None of these steps flow naturally from Canada's extant strategic culture. They are progressive steps fully in accord with the battlespace of the 21st century. Nonetheless, the jury must remain out on how successful they will be when all the other dimensions of strategic culture are brought to bear.

Strategic Culture: Technology

As an advanced industrial, now increasingly post-industrial nation, technology has always played an integral role in shaping Canada's strategic culture. However, perceptions of threat, the relatively small size of the military and constant downward pressure on defence spending have meant acquiring the most recent technology in sufficient quantity is a problem. "Rust-out" and the "commitment-capability" gap are familiar phrases in the post-WWII era. This has meant doing more with less but has also tended to avoid the worst excesses of technicism warned against by Gray.[8]

The concept of a Revolution in Military Affairs (RMA) plays a more direct and significant role in dramatically shaping a nation's strategic culture. Thus, the nuclear revolution had a dramatic effect on political-strategic doctrine in those countries that acquired these capabilities. By eschewing such weapons, itself a product of Canada's overall strategic culture, this revolution has had a more indirect and subtle effect in Canada. Similarly, for the time being at least, the movement into space has little real impact. Arguably, the revolution in information technology is having, and will continue to have, a greater impact. All associated technologies involving communications, precision weapons, sensors and weapons platforms reinforce the need for a better appreciation of joint operations and should tend to undermine the "strong single service" ideology, especially in the creation of joint, operational-level commands.

SUMMARY

When the interactive and interrelated factors contributing to Canada's strategic culture are considered holistically, what can be said by way of summary about Canada's "way of war"? First, the primacy of policy and the subordination of the military to civilian control is an enduring characteristic of strategic culture deriving primarily from geography and history. Second, although somewhat muted by history and Canadian political ideology, Canada is not a nation of peacekeepers with a military force best-suited for that role. The Canadian military was and is a combat-capable force normally supported in that role by Canadian society. That we are actually quite good at peacekeeping is more a reflection of a strong professional ethic, including responsiveness to government direction, without shirking. Third, up to the present, the Canadian military has operated almost exclusively at the tactical level, usually in a multicultural context. This has had profound implications for force structure, defence organization, military theory and higher-level doctrine. Finally, domestic influence on strategic culture has been, and will continue to be, profound. Canadians can fight but prefer to be peaceable. They are militarily capable but also quite unmilitary. Canadians want to project Canadian values throughout the international system. Canada will therefore continue to operate in an expeditionary mode.

Strategic culture is an indispensable concept when preparing for operations. It applies to allies and coalition partners, as well as the host nation (if any), and above all to the adversary. Strategic culture applies to all elements of the General System of War and Conflict – policy, strategy, operational art and tactics. Each of these sub-systems is shaped, in one way or another, by the factors of strategic culture – geography; history; ideology, religion and culture; governance and technology.

NOTES

1 Colin S. Gray, *Another Bloody Century: Future Warfare in the 21st Century*, (Oxford: Oxford University Press, 2005), 23.

2 Carl von Clausewitz, *On War*, Michael Howard and Peter Paret (eds.), (Oxford: Oxford University Press, 1976), 8.

3 Alastair Johnston, "Thinking About Strategic Culture" *International Security* 19 (4, 1994-1995), 46.

4 Cultural assumptions constitute a system of inherited conceptions expressed in symbolic forms by means of which its members communicate, perpetuate and develop their knowledge of, and attitudes towards, life. See Clifford Geertz, *The Interpretation of Cultures*, (New York: Basic Books, 1973), 89

5 John Lynn, *Battle: Combat and Culture*, (Boulder: Westview Press, 2003).

6 Williamson Murray and Mark Grimsley, "Introduction: On War," *The Making of Strat-egy: Rulers, States, and Wars*, Williamson Murray, Alvin Bernstein, and MacGregor Knox (eds.), (Cambridge: Cambridge University Press, 1996), 14.

7 Colin S. Gray, *Modern Strategy*, (London: Weidenfeld, 1999), 35.

8 As discussed on p. 29 of this chapter.

RECOMMENDED READING

Clausewitz, Carl von. *On War*. Michael Howard and Peter Paret (eds.), Oxford: Oxford University Press, 1976.

Gray, Colin S. *Modern Strategy*. London: Weidenfeld, 1999.

Gray, Colin S. *Another Bloody Century: Future Warfare in the 21st Century*. Oxford: Oxford University Press, 2005.

Johnston, Alastair. "Thinking About Strategic Culture," *International Security* 19 (4, 1994-1995).

Lynn, John A. *Battle: A History of Combat and Culture*. Boulder: Westview Press, 2003.

Murray, Williamson and Mark Grimsley. "Introduction: On War," *The Making of Strategy: Rulers, States, and Wars*. Williamson Murray, Alvin Bernstein, and MacGregor Knox (eds.), Cambridge: Cambridge University Press, 1996).

CHAPTER 4

CULTURAL FOUNDATIONS

Karen D. Davis

As an institution, the CF operates within the context of Canadian policy and strategic culture. The capacity of individuals to operate effectively within security environments, including under the auspices of the CF, is reliant upon a myriad of experiences and knowledge that are formed within Canadian cultural foundations. CF institutional culture, shaped within Canadian strategic culture, provides the context within which CF members make sense of the world. Culture provides stability, a sense of belonging and identity, and creates commitment to something larger than self-interest.[1] Military organizations, including the CF, call it military ethos. Military ethos, as an expression of culture, is essentially the social glue that holds the organization together by providing appropriate standards of behaviour, and guiding and shaping the attitudes and behaviours of members. In the case of the CF, this is presented as a doctrinal framework in *Duty With Honour: The Profession of Arms in Canada*. Importantly, culture provides a source of pride and identity for CF members, as well as a framework for understanding the world. Beginning with an overview of the domestic context, this chapter addresses some of the underlying cultural assumptions that inform the foundation of Canadian military culture and identity, and how these assumptions can be addressed to enhance the development and application of CQ.

CULTURAL INTELLIGENCE STARTS AT HOME

Experts on CQ focus on its application *across* culture; however, the basis for developing CQ is located *within* culture. The concept of mindfulness, introduced in Chapter 2, highlights the importance of knowledge not only of other cultures, but also of oneself and the way in which assumptions about one's culture and status in that culture influence perceptions and understanding of others, behaviour towards others, and the messages, both explicit and implicit, that are conveyed by one's behaviour. Clearly, CQ starts at home.

CQ builds upon *internal cultural* competencies frequently referred to as social intelligence and emotional intelligence. Social intelligence is "the ability to understand the feelings, thoughts, and behaviours of persons including one's own, in interpersonal situations and to act appropriately on that understanding."[2] Emotional intelligence (EQ) refers to "the ability to monitor one's own and others' emotions, to discriminate among them, and to use the information to guide one's thinking and actions."[3] Individuals who demonstrate high levels of social and emotional intelligence within a familiar culture may still find it difficult to understand and adjust to unfamiliar milieus.[4] However, when the skills and competencies that comprise social or emotional intelligence are compared to those contributing to CQ, there are many similarities.[5] Notwithstanding the value of social and emotional intelligence, they are not sufficient in and of themselves to the full development of CQ. CQ, as discussed in Chapter 2, is a meta-competency derived from knowledge, cognition, motivation, and behaviour integrated by mindfulness, informing one's capacity to adapt and respond in cross-cultural and unfamiliar environments.

CQ in the CF finds its foundations within the life experiences of members, including their membership in the CF itself. By the time a member of the CF deploys in an operation, the cultural foundations upon which he or she will continue to learn, interpret, and act in unfamiliar situations have already been firmly established. Values and assumptions become internalized within culture and influence the way members perceive, think, act, make decisions, and behave both on a day-to-day basis and in response to stress and unusual circumstances. Culture is learned over time and is relatively stable. In fact, it has been noted that, "[a]s a stabilizing force in human systems, culture is one of the most difficult aspects to manage in a climate of perpetual change".[6] Consequently, when faced with unfamiliar situations that were not addressed in pre-deployment cultural awareness training or situations that just do not make any sense, those values and understandings that have been an integral part of an individual's experience will guide his/her response.

The discussion which follows highlights the ways in which our cultural orientations contribute to, and have the potential to impair, our CQ. Key challenges in this regard include developing an awareness of the strengths and biases of our own culture and recognizing when and how our own cultural orientations are influencing our beliefs and behaviours, as well as influencing our ability to interpret and understand the beliefs and behaviours of others.[7]

CANADIAN VALUES AND IDENTITY

To a significant extent, cultural foundations and identity are formed well before an individual joins the CF. Most Canadians take great pride in being Canadian; however, many are also likely to identify first with a specific region or city.[8] Canadian identities are grounded within ethnic, regional and national values and practices – both espoused and real. Canada as a nation is not one culture, but an amalgamation of many cultures that share a fundamental commitment to the democratic ideal; peace, order and good government; and the rule of law. Canadian values are expressed in legislation such as the *Canadian Charter of Rights and Freedoms*, the *Criminal Code of Canada*, and acts of parliament such as the *Canadian Human Rights Act*, the *Employment Equity Act*, and the *Multiculturalism Act*.

Basic Canadian values can be quite similar to those of many other cultures, communities, and nations, but also very different. Fundamentally, Canadians value the right of the individual to make lifestyle and values-based choices, respect the equal rights of all women and men, and believe in freedom of expression. Over 75 per cent of Canadians believe that they have a great deal of freedom of choice and control over their lives;[9] a majority of Canadians have confidence in the police and the legal system,[10] approve of the human rights movement[11] and believe that God is very important in their lives.[12] Opinions and beliefs such as these reflect the shared underlying values and experiences that inform how Canadians interpret and view the world, and provide an ideological reference point for Canadian culture.

Culture and cultural identities within society become increasingly complex and multiple as the society becomes more diverse. Individuals may embody a cultural consciousness that is not simply dualistic (e.g. African-Canadian), but a 'poly consciousness' reflecting ethnicity, region or origin, language, religion and so on.[13] Culture is an ambiguous and complex concept that is used to try to categorize and make sense of structured and virtual groups of individuals, organizations, communities, and nations. Increasingly, organizations are comprised of multiple and complex cultural influences with varying relationships and implications for the organization. In fact, the inclusion of various cultural influences in the CF is integral to its identity as a Canadian institution as Canada becomes more diverse.

CANADIAN FORCES VALUES AND IDENTITY

In spite of the diversity that is represented within the CF, it is important to acknowledge the roots of CF culture and the ways in which these roots impact the CF perception of the world. As a Canadian institution, the CF strives to represent Canadian values and foundations at home and abroad. CF members experience intense socialization into the military through training, professional development, employment, ritual, and tradition. CF members are encouraged to understand the uniqueness of their role within Canadian society. *Duty With Honour: The Profession of Arms in Canada* not only describes the CF member as unique, but also emphasizes the obligations and responsibilities that go along with this role, on behalf of Canada:

> In volunteering for military service, members of the Canadian Forces accept as part of their Duty a unique and distinct identity within Canadian society, yet in subtle ways they are apart from it. By embracing the military ethos, they accept obligations and responsibilities that no other Canadian citizen has. As members of the profession of arms, they are acutely aware of the special trust placed in them by the people of

> Canada. They accept this trust, and in accord with their ethos, strive for excellence within their areas of expertise. In brief, they reach for the highest standards of professionalism.[14]

This characterization is integral to ensuring that the Canadian military professional takes his/her role seriously and thus fulfills the essential obligations that go with the uniform. As a theoretical ideal, the doctrine that guides the profession encourages CF members to take exceptional pride in their roles and in Canada as a nation. Duty to Canada; loyalty to Canada and comrades; the integrity to adhere to high ethical standards; and the physical and moral courage to disregard personal loss and make difficult choices, are core military values familiar to all CF members.[15]

Strong identity, pride, and cultural self-esteem are critical to effective operations. It is imperative that CF members have confidence in Canada and the values that it represents; however, excessive pride can manifest itself as *ethnocentrism*, thus representing a potential inhibitor to the development of awareness and understanding of potential partners and belligerents. *Ethnocentrism* refers to:

> *Attitude or prejudice based upon the belief that one's own culture is considered to be racially, morally and culturally of greater value or significance than that of others; involves an incapacity to acknowledge that cultural differentiation does not imply the inferiority of those groups who are ethnically or culturally distinct from one's own.*[16]

While it is natural to use taken for granted understandings as knowledge against which to compare other people and cultures, it can also create significant barriers to understanding difference. Understanding and respecting difference opens the door to further understanding, effective decision-making, and culturally appropriate responses, not least of which might be the capacity to anticipate the actions of a belligerent. An open mind is an important contributor to CQ.

Understanding cultural difference implies due regard for that which is different; however, it does not necessarily mean the acceptance of all difference as an attribute rooted in cultural beliefs and values. In addition, heeding cultural difference is not the same as treating those differences with deferential esteem. Also, it is important to make distinctions between those phenomena that are shared across cultures from those things that represent meaningful cultural difference. *Cultural relativism* refers to the extent to which differences are accepted as a cultural or "local" matter. There are numerous debates surrounding the relationship between cultural differences and universal rights and wrongs. In a 2007 issue of *Anthropological Quarterly*, there are several viewpoint essays addressing the concept of cultural relativism. For example, according to anthropologist, Robert C. Ulin:

> Cultural relativism is among the most misunderstood yet socially charged concepts associated with anthropology today. While American cultural anthropologists have utilized cultural relativism as a pedagogical and sometimes political medium to challenge ethnocentric western views and cultural practices and to promote an appreciation of cultural diversity, ethicists, philosophers and the general public have all too often embraced a view of cultural relativism that ostensibly allows repugnant customs and social practices to go unchallenged.[17]

Culture is not clearly this way or that way, but a complex mixture of good, bad, and indifferent universal practices and culturally relative phenomena. The interpretation of cultural practices and values can be limited by ethnocentrism on the one hand, and cultural relativism at the other extreme. This inherent ambiguity and complexity of culture, cultural motivation, and apparent cultural practices underscore the importance of CQ as a critical meta-competency for the CF.

CULTURAL INTELLIGENCE AND DIVERSITY IN THE CANADIAN FORCES

Today, the CF is much better positioned to operate in diverse environments and domains, considering current levels of CQ compared to the CQ that existed within the organization 10, 20, 30 and more years ago. As a result of the social evolution of Canadian society and subsequently the CF, CF leaders are less able to rely on authoritative leadership based upon positional authority than in the past. Alternatively, CF doctrine encourages transformational leadership styles which rely on shared core values, mutual commitment, and trust, while emphasizing that today's leaders "must continue to strive for a common identity and teamwork within a more varied and complicated human resources landscape."[18] In building individual commitment and teamwork, the CF has placed increasing emphasis on the development of strong interpersonal skills such as building relationships, valuing people of different cultures, listening and observation, coping with ambiguity, translating complex information, taking action and initiative, managing others, adaptability/flexibility, and managing stress.

In recent decades, experience with both gender integration and diversity has provided important opportunities for the CF to develop CQ through the increasing ability of leaders in the organization to understand, respect, and negotiate a variety of differences.[19] For example, as the military works more frequently alongside different networks, agencies, and organizations that include women, the capacity to negotiate culture along sex and gender[20] lines is essential to CQ. Harassment prevention and resolution, alternative dispute resolution, diversity, and ethics are integrated aspects of the professional development of CF members. The findings of a 1999 diversity survey suggest considerable improvement in CF member attitudes and behaviours towards cultural diversity since the initial administration of the survey in 1996, thus indicating increased acceptance of a diverse society and willingness to accept and respect others who are culturally or racially different.[21]

The CF has adopted an approach to gender and diversity which posits that if behaviour changes first, attitudes will follow over the longer term. The CF tip book, *One Team, Many Faces: Employment Equity and Diversity Leadership*, opens with a section entitled "Change Behaviours First, Attitudes Later," which advises that success depends upon "leaders carrying out their responsibilities in accordance with the law and the expectations of Canadians, *regardless of personal views*."[22] If one changes their behaviour, attitude change will follow. This approach is necessary to communicate the organization's commitment to achieving legislated objectives; however, it does present limitations for gender, diversity and CQ among CF members. Behaviour that is coercively based upon the legality of equitable treatment, in the absence of understanding and willingness to understand difference, is unlikely to result in culturally effective and adaptable problem-solving and decision-making. In a culture that relies disproportionately on shared values and experience, and specific cultural knowledge, there is not enough trust, resilience, and adaptability to effectively understand and accommodate difference in ways that satisfy both equality and operational success on a consistent basis.

Similarly, a homogenous military culture that relies on pre-deployment cultural information to raise awareness and effectiveness will experience a limited capacity to develop effective CQ and apply effective *understanding* when confronted with unique experiences and circumstances. Knowledge is a significant contributor to the cognitive process, but knowledge alone is not sufficient to develop optimum effectiveness in cross-cultural domains. Behaviour is also an important contributor to CQ, but behaviour alone does reflect CQ. The analyses of the 2005 *Diversity Climate Survey* in the CF found that reported overt behaviours were, overall, more positive than the reported attitudes toward gender and diversity.[23] These findings confirm at the very least that attitudes toward gender and diversity are lagging behind behaviour in terms of desired organizational responses, and as mentioned earlier, highlight limitations in terms of developing CQ.

Regardless of the challenges, effective leadership of diverse teams is expected of all CF leaders. The CDS Guidance to Commanding Officers includes a section on "Human Rights and Diversity," that begins by establishing a clear relationship between operational effectiveness and the treatment of all members with dignity and respect:

> Our men and women are the cornerstones of operational effectiveness. Sophisticated weapon systems, armoured vehicles, aircraft and ships are of little value without a dedicated team of professional sailors, soldiers and air persons. Our people must know they will be treated fairly and

> with dignity and respect throughout their careers. They have a right to expect from their leadership an environment that is free from discrimination and harassment in an organization whose employment practices are fair and equitable.[24]

The leadership skills that contribute to leadership and diversity within the CF are complementary, and in some cases identical to, the foundational skills of CQ. Most experienced leaders will recognize the dilemma of making a decision to ensure fairness for one individual or group of members, only to create resentment among those who do not share the same characteristics and thus do not perceive that they are in a position to benefit from the decision. Leaders also understand the importance of shared experience in developing cohesive and productive teams and as such will frequently make decisions that are applied equally to all. Regardless of difference, it is often necessary to treat all team members the same in an organization such as the CF. However, this approach is too frequently adopted as the default position that is considered to be reflective of *gender-neutral* or *colour-blind* perspectives, with little or no awareness of the differences that have been rendered invisible or problematic from the perspective of dominant cultural values and understanding. Culturally intelligent leaders are able to negotiate and satisfy these conflicting demands by seeking and understanding meaningful differences, beyond the superficial and visible, in ways that enhance, rather than impair, mission success.

As discussed in Chapter 2, mindfulness is a key component of CQ. Mindfulness highlights the importance of knowledge of other cultures, as well as knowledge of oneself and the way in which assumptions about one's own culture and status in that culture influence perceptions and understanding of others, behaviour toward others, and the messages, both explicit and implicit, that are conveyed by behaviour. CQ experts, David C. Thomas and Kerr Inkson, suggest that our multicultural society does present numerous opportunities to develop CQ and as such identify several basic "rules of engagement"[25] to keep in mind when interacting with those who are culturally different:

- build relationships;
- expect differences in others, see different behaviour as novel, and suspend evaluation;
- be attentive to behavioural cues, their possible interpretation, and the likely effect of your behaviour on others;
- become knowledgeable about one's own culture and background, its biases and idiosyncrasies, and the way this is unconsciously reflected in your own perceptions and behaviour;

- adapt behaviour in ways that are comfortable and believed to be appropriate for the situation;
- be mindful of responses to behavioural adaptation;
- experiment with methods of adapting intuitively to new situations, and use these experiments to build a comfort level in acquiring a repertoire of new behaviour; and
- practice new behaviours that work until their production becomes automatic.

Thomas and Inkson caution that superficial interactions, lacking depth and intensity, are often not enough to push an individual outside of their comfort zone. Thus, those seeking to become more culturally intelligent need to find challenges that move them outside of familiar and comfortable domains.[26] However, CF members do not have to leave Canada to experience organizational and community cultures that they do not understand.

UNDERSTANDING MILITARY CULTURE

CQ has applications across organizations, networks, groups, and communities across domestic, international, host nation and enemy domains. The principles of CF transformation, promulgated by the CDS in 2005, give precedence to CF culture over navy, army, and air force affiliations, as well as placing significant priority on an integrated Regular, Reserve and Civilian Public Service. All members of the CF, regardless of unit or environmental affiliation, are expected to share a commitment to core Canadian and military values; however strong single-service cultures do exist and often impede progress in this area of transformation. Environmental service, ship, squadron, and regimental cultures promote loyalty and identity among their members through rituals and traditional practices. Differences across cultures grounded in similar values can be subtle and difficult to understand; however, it is important to find ways to understand and negotiate differences to facilitate operational success.

Comprehensive academic studies provide various and useful tools for accessing cultural information and providing frameworks for understanding the CF and defence and security cultures in Canada, as well as domestic and international military and civilian organizational cultures with various goals and imperatives. Donna Winslow's socio-cultural study of the Canadian Airborne Regiment in Somalia is perhaps the most comprehensive cultural study of a CF unit ever conducted. Employing a social anthropological perspective, this cultural inquiry was conducted using in-depth interviews, focus groups, visual records such as photos, personal records such as letters written home, first-hand accounts written by Somali journalists, and official documents and records.[27] Importantly, her findings emphasize that exaggerated unit loyalty and identity can result in disrespect for, and isolation from, other military units, dysfunctional behaviour, and loss of ability to relate to anyone outside the group.[28] The end result is not only an ever-increasing alienation from Canadian society, but also a severely limited ability to effectively operate in cross-cultural environments of any kind.

At the other end of the spectrum, Canadian historian Allan English employed secondary sources to develop an analysis of Canadian military culture within the context of Canadian society, including an examination of the roots of Canadian military culture, the contributions of the air, land and sea subcultures, ethos, and professionalism.[29] English relies heavily on Edgar H. Schein's leader-driven model of culture, which has informed CF leadership doctrine.[30] Schein's model is useful in providing important clues about what is important in an organization and, therefore, motivates its members.[31] However, English cautions that while organizational culture theorists will frequently portray culture as dominant and cohesive, it is also important to consider frameworks that facilitate understanding of differentiation, inconsistency, ambiguity, and conflict within organizations and networks.[32] Consequently, it is important to consider alternative approaches that place greater priority on the analysis of relational aspects that consider, for example, internal conflict within organizations and cultures.[33]

Military anthropologist, Charles Kirke, notes that it can be misleading to treat one human group the same as another; he suggests that there is good evidence to support cross-inference when organizations share key social and structural characteristics. He posits, for example, that his model of military organizational culture can be applied to develop understanding of other military cultures, such as the CF, that share key characteristics with the British army model for which the following four social structure categories have been developed:

- ***formal command structure:*** structure through which a member at the bottom receives orders from the person at the top;
- ***informal structure:*** unwritten conventions of behaviour in the absence of formal constraints;
- ***loyalty/identity structure:*** series of different sized groups which are defined by opposition to and contrast with other groups of equal status in the formal command structure; and
- ***functional structure:*** attitudes, feelings, and expectations connected with being 'soldierly' and properly carrying out 'soldierly' activity.[34]

Groups of key characteristics such as these provide examples of frameworks that can be applied to enhance understanding the capacity of military units, including one's own, to operate effectively in domestic and international cross-cultural environments.

DEPLOYING CULTURAL INTELLIGENCE

The case of CF domestic aid to civil power at Oka in 1990 (presented at Appendix C) provides an instructive example of the importance of CQ when working with diverse communities in Canada. In the case of Oka, the CF was largely successful, especially when compared to the experience of the Sûreté du Québec, whose lack of cultural knowledge and aggressive behaviour resulted in significant difficulties. The CF was able to declare mission success; however, there are standing resentments that will linger for years to come between the Mohawk community and the CF as a result of Oka. Regardless of the degree of success experienced by the CF, it is worth considering how things might have been different if the CF had operated at an even higher level of CQ than was portrayed to the public during and in particular after Oka.

Mission success is also increasingly dependent upon the CF's working relationships with other Canadian government departments and agencies, in particular the Department of Foreign Affairs and International Trade (DFAIT) and the Canadian International Development Agency (CIDA), in achieving integrated defence, diplomacy, and development effects. A broader whole of government approach spanning numerous federal departments, as well as provincial and municipal governments in domestic and international operations also demands the capacity to work effectively with organizational cultures which reflect different priorities and processes than those driving CF activity and decision-making.

Andy Tamas, a capacity development specialist and former member of an integrated military-civilian Strategic Advisory Team (SAT) in Afghanistan, has identified various ways in which military assumptions and priorities can create barriers to understanding other cultures and thus undermine progress. For example, in one situation a military member expressed extreme frustration with the slow progress in work with local civilians, asserting that military officers are trained to *act* and *achieve* results. As noted by Tamas in the quote below, it is important to understand how culture is influencing behaviour in particular situations. That is, it is important to,

> ...sufficiently appreciate the full implications of the fact that all behaviour is intentional and appears rational to the person or group at the

> time. To help a system move it is essential to figure out the motivation structure behind what is being observed and seeing what can be done to shift internal patterns so the system progresses on its own.[35]

In a case such as this, a culturally intelligent approach would allow an observer to see the progress that is being made even if that progress is different than anticipated, or adapt behaviour to facilitate greater progress.

Although there are obvious differences in organizational missions and mandates, the lines between humanitarian and security operations, for example, are increasingly blurred. In addition, a variety of communities and organizations are dependent upon one another in various ways to achieve their respective goals. CQ provides the key to breaking down barriers to enhance mission success. The contrasting perceptions of military and humanitarian organizations presented in Table 4-1, below, provide a few examples of differences that have become evident to military sociologists and illustrate the importance of CQ in overcoming both the real and perceived differences in operating cultures and priorities. When this type of *knowledge* is accepted as truth, in the absence of further effort to understand the culture of the other, it is unlikely that effective partnerships will be developed to ensure that all parties achieve their shared and individual goals.

NGO perceptions of military	**Military perceptions of NGOs**
Boys with toys	Non-guided organizations
Rigid	Children of the sixties
Authoritarian	Flaky do-gooders
Conservative	Permissive
Impatient	Unpunctual
Arrogant	Obstructionist
Civilian phobic	Anarchic
Homophobic	Undisciplined
Excessively security-conscious	Self-righteous

TABLE 4-1: NON-GOVERNMENTAL ORGANIZATIONS (NGO) AND MILITARY PERCEPTIONS[36]

To complicate matters, cultural motivations and responses will change to adapt to various circumstances. Underlying and largely invisible cultural assumptions, values, and world views remain fairly static; however, when faced with crisis, cultural responses cannot be consistently predicted based upon rules or stereotypes. Cultural awareness training, immediately prior to deployment, provides knowledge of the anticipated situation that will be faced; however, no amount of knowledge will fully prepare CF members for every situation that will be encountered.

SUMMARY

CF members are called on to perform a range of missions, not least of which is to promote Canadian values and ideals through their actions at home and abroad. Mission success is dependent upon the capacity of CF members to learn, understand, and adapt to cross-cultural environments. In recent decades, the social values of the CF have evolved to reflect those of Canadian society. CF leaders have been challenged by a range of social and political challenges stemming from legislation and the changing security environment. As a result, CF members are encouraged to develop those social competencies that contribute to leadership, and also provide a strong foundation for the development of CQ. Importantly, the identity of a CF member is based upon a sense of national pride and patriotism, and military duty, loyalty, integrity, and courage. In addition to negotiating the challenges of unfamiliar cultures, complex systems and relationships, and difficult missions, CF members must also demonstrate moral responsibility. Chapter 5 provides insight into the relationship between identity development, CQ and decision-making. This includes those decisions which reflect moral judgements derived from the way in which people make sense of the world around them.

NOTES

1 Stephen P. Robbins and Nancy Langton, *Organizational Behaviour: Concepts, Controversies, Applications* (3rd Canadian Edition). (Toronto: Prentice Hall, 2003), 349.

2 H. Marlowe, "Social Intelligence: Evidence for multidimensionality and construct independence," *Journal of Educational Psychology,* 78 (1986) 52-58 cited in P. Christopher Earley and Soon Ang, *Cultural Intelligence: Individual Interactions Across Cultures*. (Stanford, California: Stanford Business Books, 2003), 46.

3 P. Salovey and J. D. Mayer, "Emotional Intelligence," *Imagination, Cognition, and Personality* 9 (1990), 185-211 cited in Earley and Ang, *Cultural Intelligence*, 47.

4 Earley and Ang, *Cultural Intelligence*, 4.

5 Kimberly-Anne Ford and Karen D. Davis, *Cultural Intelligence, Emotional Intelligence and Canadian Forces Leader Development* (Kingston, ON: Canadian Forces Leadership Institute, 2007), CFLI Technical Memorandum 2007- 01.

6 Edgar H. Schein, *Organizational Culture and Leadership*. (San Francisco: Jossey-Bass, 1992), xiv.

7 See, for example, David C. Thomas and Kerr Inkson, *Cultural Intelligence: People Skills for Global Business*. (San Francisco: Berrett-Koehler Publishers, Inc., 2003), 10-11.

8 Ronald Inglehart, Miguel Basañez and Alejandro Moreno, *Human Values and Beliefs: A Cross-Cultural Sourcebook*. (Ann Arbor, The University of Michigan Press, 2005). Sixty-one per cent of Canadian participants in a cross-cultural values survey indicated that they were

very proud to be Canadian, V322 ; 40 per cent identified first of all with Canada, V320; 31 per cent with the local area that they were from, V320; and an additional 13 per cent identified with a geographical area beyond nation (e.g. continent, world), V320.

9 Inglehart, Basañez and Moreno, *Human Values and Beliefs*. V95. Highest percentage is Finland at 79 per cent; lowest Bulgaria at 28 per cent. Countries included in the sample and thus comprise *worldwide* are in order of Gross National Product (GNP) per capita: India, Nigeria, China, Romania, Turkey, Poland, Bulgaria, Chile, Czechoslovakia, South Africa, Lithuania, Hungary, Argentina, Brazil, Mexico, Belarus, Russia, Moscow, Latvia, Estonia, Portugal, South Korea, Ireland, North Ireland, Slovenia, Spain, East Germany, Britain, Italy, Netherlands, Belgium, Austria, France, Canada, United States, Iceland, West Germany, Denmark, Finland, Norway, Sweden, Japan, and Switzerland.

10 Ibid. 84 per cent of Canadians expressed a *great deal* or *quite a lot* of confidence in the police compared to 55 per cent worldwide (highest Denmark 89 per cent; lowest Estonia 19 per cent), V278; and 54 per cent expressed confidence in the legal system compared to 52 per cent worldwide, V275.

11 Ibid. 58 per cent compared to 62 per cent worldwide (highest Nigeria 86 per cent; lowest Estonia 27 per cent), V293.

12 Ibid. 62 per cent compared to 50 per cent worldwide (highest Nigeria 98 per cent; lowest China 3 per cent), V176.

13 George Elliott Clarke. 1998. "Contesting a Model Blackness: A Meditation on African-Canadian Americanism, or The Structures of African Canadianite." *Essays in Canadian Writing*, 63 cited in Liane Curtis, Dipti Gupta and Will Straw. 2001. "Culture and Identity: Ideas and Overviews" paper commissioned by the Department of Canadian Heritage.

14 Canada, *The Profession of Arms in Canada*. (Kingston, ON: Canadian Forces Leadership Institute, Canadian Defence Academy, 2003), 77.

15 Ibid., 30-31.

16 David Jary & Julia Jary, *Web-linked Dictionary: Sociology,* (Glasgow: Harper Collins Publishers, 2000), 193.

17 Robert C. Ulin, "Revisiting cultural relativism: old prospects for a new cultural critique," *Anthropological Quarterly*, 80 (3, Summer, 2007).

18 Canada, *Leadership in the Canadian Forces: Conceptual Foundations* (Kingston, ON: Canadian Defence Academy, Canadian Forces Leadership Institute, 2005), xv. Available online at: <http://www.cda-acd.forces.gc.ca>.

19 See Karen D. Davis, "Women, Gender, and Cultural Intelligence in the Canadian Forces," paper presented at *Defending Democracy: Accommodating Diversity in the Security Sector* conference, Kingston, ON: Queen's University, 1-3 November 2007.

20 Sex and gender are commonly used interchangeably; however, sex refers only to biological differences between women and men, while gender refers to patterns of masculinity and femininity, which influence expectations, roles, behaviours, etc. of women and men. Within the academic literature, gender has been commonly understood as a socially constructed concept for more than 25 years, and contemporary understandings of gender are

rooted within the status of women relative to men in society. See for example, Mark Hussey, "Preface," *Masculinities: Interdisciplinary Readings*, (Upper Saddle River, New Jersey: Prentice Hall, 2003), iii; Status of Women Canada, *Gender-Based Analysis: A guide for policy-making*, (Ottawa, ON: Status of Women Canada, 1996), and Karen D. Davis, "Gender", *The Military Leadership Handbook*. Bernd Horn and Robert W. Walker (eds) (Toronto: Dundurn Press, 2008), 309-319.

21 H. Pike, R.N. MacLennan, and N.M. Perron, *Canadian Forces Diversity Climate Project: 1999 Survey Administration.* (National Defence, Canada: Director Human Resources Research and Evaluation, 2000).

22 Canada. Department of National Defence. 2003. *One Team, Many Faces: Employment Equity and Diversity Leadership*, based on the book *Employment Equity and Diversity Management – A Tip Book* by Lauren Nancoo, p. 12.

23 The results of the *Diversity Climate Survey* administered to CF members in 2005 indicates that on a scale of one to five, with five representing the most favourable attitudes toward gender integration, the average Regular and Reserve Force responses were 3.47 and 3.57. Response average was more favourable among the senior officers in the Regular Force and among the Junior Officers in the Reserve Force. NeoSexist and NeoRacist scales administered in the same survey produced similar results. That is, on a scale of one to seven with one indicating the lowest level of subtle racist or sexist attitudes, CF average responses were slightly toward the positive end of the scale ranging between 3.66 (Regular Force NCMs on NeoRacist Scale) to 3.07 (Reserve Force Junior Officers on NeoSexism Scale). Overall, average responses were slightly less negative on the NeoSexist Scale than on the NeoRacism scale. S. Urban and I. Goldenberg. 2006. *Canadian Forces Diversity Climate Project 2005 Administration*, (Ottawa: Department of National Defence), Director Personnel Applied Research Sponsor Research Report 2006-08.

24 Canada, *CDS [Chief of the Defence Staff] Guidance to Commanding Officers*. Promulgated since 2003. Web-based access since 2005: <http://cda.mil.ca/cdsguidance/engraph/home_e.asp> /<http://cda.mil.ca/cdsguidance/frgraph/home_f.asp>

25 David C. Thomas and Kerr Inkson. *Cultural Intelligence: People Skills for Global Business*. (San Francisco: Berrett-Koehler Publishers, Inc., 2003) p. 78.

26 Ibid., 75.

27 Donna Winslow, *The Canadian Airborne Regiment in Somalia: A Socio-cultural Inquiry*. (Ottawa, ON: Ministry of Public Works and Government Services, 1997).

28 Donna Winslow, *The Canadian Airborne Regiment in Somalia*, 263-264.

29 Allan D. English, *Understanding Military Culture: A Canadian Perspective* (Montreal & Kingston: McGill-Queen's University Press, 2004).

30 See Canada, *Leadership in the Canadian Forces: Conceptual Foundations* (Canadian Defence Academy, Canadian Forces Leadership Institute, 2005), 116; and Canada, *Leadership in the Canadian Forces: Leading the Institution* (Canadian Defence Academy, Canadian Forces Leadership Institute, 2007), 13-14.

31 Edgar H. Schein, *Organizational Culture and Leadership*, second edition. (San Francisco: Jossey-Bass, 1992), 231.

32 Allan D. English, *Understanding Military Culture*, 20.

33 For example, Actor Network Theory (ANT) considers the stable and enduring aspects within institutional culture to be relational effects that mask "an ongoing and dynamic internal struggle between competing actor networks. See Thomas B. Lawrence and Roy Suddaby, "Institutions and Institutional Work," *The Sage Handbook of Organization Studies* (2nd Edition). Stewart R. Clegg, Cynthia Hardy, Thomas B. Lawrence and Walter R. Nord (eds.), (London: Sage Publications, 2006), 242.

34 Charles Kirke, "The Organizational Cultural Approach to Leadership: 'Social Structures' – A Tool for Analysis and a Way Ahead," *Dimensions of Military Leadership*. Allister MacIntyre and Karen D. Davis (eds.) Kingston, ON: Canadian Defence Academy Press, 2006), 288-291.

35 Andy Tamas, *Warriors and Nation Builders: Development and the Military in Afghanistan*. (Kingston, ON: Canadian Defence Academy Press, 2009), 57.

36 Donna J. Winslow, Lindy Heinecken, and Joseph L. Soeters, "Diversity in the Armed Forces," *Handbook of the Sociology of the Military*. Giuseppe Caforio (ed.) (New York: Kluwer Academic/Plenum Publishers, 2003), 307.

RECOMMENDED READING

Davis, Karen D., "Women, Gender, and Cultural Intelligence in the Canadian Forces," paper presented at *Defending Democracy: Diversity in the Security Sector* conference, Kingston, ON: Queen's University, November 1-3, 2007.

Earley, P. Christopher and Soon Ang. *Cultural Intelligence: Individual Interactions across Cultures*. Stanford, California: Stanford Business Books, 2003.

Allan D. English. *Understanding Military Culture: A Canadian Perspective*. Montreal & Kingston: McGill-Queen's University Press, 2004.

Andy Tamas. *Warriors and Nation Builders: Development and the Military in Afghanistan*. Kingston, ON: Canadian Defence Academy Press, 2009.

Robert C. Ulin. "Revisiting cultural relativism: old prospects for a new cultural critique," *Anthropological Quarterly*, 80 (3, Summer, 2007).

Winslow, Donna J., Lindy Heinecken, and Joseph L. Soeters. "Diversity in the Armed Forces," *Handbook of the Sociology of the Military*. Giuseppe Caforio (ed.) New York: Kluwer Academic/Plenum Publishers, 2003.

CHAPTER 5

CULTURAL INTELLIGENCE AND IDENTITY DEVELOPMENT

Dr. Daniel Lagacé-Roy and Justin C. Wright

The aim of this chapter is to explore the relationship between the concepts of CQ and identity development. More specifically, this chapter looks at personal identity development in relation to the meaning of one's own cultural values, beliefs, and expectations, and to the meaning of unfamiliar ones. The relationship between these two "meanings" is sometimes perceived as one "right" and the other "wrong," thus our culturally developed values and practices can be understood to be what drives our particular way of moral reasoning. The "role" of identity in this relationship is to *make meaning* of the foundation on which one's assumptions and understandings are predicated.

This chapter contends that the development of identity deals with cultural values and beliefs. Furthermore, it suggests that within the dynamic dimension of identity (specifically, the individual's process of *meaning-making*) lies the cognitive capacity to reflect on issues such as "conflict of values," "prejudices" and "subjectivity/objectivity", and to articulate concepts such as ethnocentrism and relativism. Therefore, it is argued that this cognitive capacity requires an understanding of personal identity development that focuses on one's culturally-informed ability to *make meaning* of one's role as, for example, a Canadian soldier in a foreign country who is defending Canadian values and *how* these values are perceived and interpreted by the *other(s)*. Conversely, this chapter discusses the question of *why* these values should be adopted by *other(s)* and/or *how* and *why* others' values have to be taken into account and therefore, defended.

This chapter uses the five stages of Robert Kegan's identity development framework as the basis for the process for *making meaning* of one's culture (or other cultures).[1] For the purpose of this chapter, an applied example is introduced; in the context of Canada's mission in Afghanistan, the example chosen addresses the intricacies of the Islamic world and more particularly the Islamic faith. The purpose of this example is not to advance a critical view of the Islamic world and faith. On the contrary, it is to provide a realistic and contemporary example of how one's identity (i.e., Canadian soldier), shaped by cultural values and beliefs (i.e., personal and professional), plays a significant role in perceiving, interpreting, evaluating and understanding the Islamic world and faith.[2]

It is within the cognitive dimension of CQ (introduced in Chapter 2) that one's identity acquires its full meaning; with this cognitive focus in mind, this chapter is divided into four parts as follows:

The first part re-visits the definition of CQ with an emphasis on the relation between CQ and the concept of *meaning-making*. The definition introduces important capacities that are fundamental when discussing and applying CQ.

The second part of the chapter introduces the role of cultural values and beliefs in shaping individual personal and professional identity and in guiding the application of CQ. To that effect, the individual's values and belief system – the key component of their identity development – determines what is required and/or appropriate as outcomes (i.e. evaluation, behaviour) in different cultural interactions.

The third part provides a definition of the meaning of identity along with the stages of identity development presented by Kegan in his book *The Evolving Self: Problem and Process and Human Development*.[3] He builds on the work of developmental psychologist Jean Piaget,[4] (i.e., systems of thought about the physical world) as well as the work of psychologist Lawrence Kohlberg[5] on moral reasoning (i.e., personal construction of the social world). Following the rationale of these two researchers on *how* children perceive and construct their personal and social world, Kegan elaborates a framework that speaks of a person's involvement in *making meaning* of his/her surrounding world. According to this author, an individual *makes meaning* of his/her surrounding world through the structuring of a personal and/or social and/or professional identity. For example, in the case of military members, this is exemplified by the values and practices of the profession of arms, which contribute to the professional identity and the demands of their organization. *Making meaning* of their professional identity, as we will see, has a direct impact on the way military members will interact with the *other(s)*, such as people of different cultures.

The fourth part concludes with challenges for military members. Some claim that a person's identity development is part of a continuum that includes an important shift from a manner of "doing" to a manner of "being." This shift challenges military members to internalize the values, beliefs, and expectations that underpin the military profession, and thus internalize these values and regulate their own actions and behaviours. More importantly, this self-regulation meets what the philosopher Michel Foucault calls "a way of being and of behaviour."[6]

CULTURAL INTELLIGENCE AND MEANING-MAKING

The cognitive dimension of CQ concerns the capability of individuals to reformulate conceptions of self and others and create new conceptualizations of how to function and operate within a new culture.[7] Within the framework of this broad reference to cognitive skills "a person's mental representation of [his/]her own knowledge and experience, social identity, and social roles"[8] is represented and enhanced. The cognitive dimension is an important aspect of CQ because it is the domain that requires the person to "bracket" or set aside pre-existing conceptualizations of *how* and *why* people function as they do.[9] It emphasizes the need for a person to incorporate, understand, and reason new information by creating a "proper mapping"[10] of *oneself* first and of *other(s)* second. This "proper mapping" is the ability to draw upon necessary knowledge for establishing principles for future interactions. This knowledge includes, but is not limited to, a broad anthropological view on social and religious systems, and political and economic institutions.

It is important to briefly address here the concepts of *oneself* (understood as "*soi-même*")[11] and of *other(s)*. These two philosophical concepts are often overlooked and taken for granted when the subject of culture is addressed, for the simple reason that the subject of culture is often viewed as a reality that concerns an all-inclusive "object." While this inclusive approach is questionable,[12] it does raise the inevitable question of *who* the "subject" of a given culture is. It is argued that the recognition of *oneself* and the *other(s)* is the terrain in which CQ (and meaning-making) takes place. Therefore, the dualistic or simplistic view of *oneself* in opposition to the *other* is challenged. The *other* is considered as a different reality such as a specific culture,[13] or the *other* viewed as the "same" such as a person sharing the same universal experience of humanity. More importantly and according to philosopher Maurice Bellet, the *other*: "…is anyone that we are concerned about."[14]

At this point, it seems appropriate to mention identity development – and values and beliefs – because it is the cognitive dimension that underpins the development of someone's identity. Furthermore, the question of values and beliefs is also addressed in the cognitive dimension. While values and beliefs represent only one of numerous features of CQ, it is an important one.[15] As discussed in the next section, the intent in focusing on values and beliefs is to create and raise awareness of the fact that cultures and people are for the most part defined by them.

VALUES AND IDENTITY

In Chapter 4, both Canadian and CF cultural values were defined and discussed. It was noted that personal and professional values stem from the culture in which one lives and thus the way in which one views the world is largely influenced by one's cultural background. In this chapter, it is necessary to revisit the topic of cultural values and to make the explicit link between those values and the identity development framework.

As was apparent in Chapter 4, the concept of "values" is typically viewed in a very generic sense: it is the result of evaluations and a concept of the desirable. Importantly, the definition of values provided by the *Oxford Dictionary of Philosophy* captures the relationship to the process of decision-making: "[t]o acknowledge some feature of things as a value is to take it into account in decision-making, or in other words to be inclined to advance it as a consideration in influencing choice and guiding oneself and others."[16]

The importance of such definition is to confirm that values, through influence and guidance, structure and shape someone's conduct. According to sociologist Michel Maffesoli, such definition goes beyond the simple naming of *what values are* (e.g., subjection to desire) and suggesting *for what they serve*, which is to articulate a mode of being.[17] Moreover, it is argued that this expression of a mode of being cannot only be conceptualized, but needs to be lived in the context of someone's personal identity and cultural endeavour. In other words, the conceptualization of one's values is a crucial component of individual identity, and thus how one lives within the context of his/her culture where "the life of values is inseparable from the conditions of socialization of each human subject."[18]

In the context of identity development and CQ, there are two areas where values are usually well defined: the professional level and the personal level. While these two levels are introduced below, it is important to keep in mind that the purpose of addressing these values is to highlight the need to have a deeper understanding of their influence on the construction of a person's identity.

First Level: The Military Profession

Again, returning to Chapter 4, it was noted that the significance of values at the professional level for a military member is clearly defined in doctrinal documents such as *Duty with Honour: The Profession of Arms in Canada* and *Leadership in the Canadian Forces: Conceptual Foundations*.[19] In *Duty with Honour,* our cultural values concerning the military are organized in three groupings: Canadian military values (e.g., duty, loyalty, integrity, and courage); Canadian values that are defended by each Canadian (e.g., equal rights, freedom of choice and expression, respect for the dignity of all people, human security, obedience and support of lawful authority); and beliefs and expectations of military service that are endorsed by each military member (e.g., unlimited liability, fighting spirit, discipline and teamwork).[20]

By stating that the Canadian military values are congruent with the values of Canadian society, *Duty with Honour* visibly advances that these values, along with beliefs and expectations, are incorporated within the profession of arms and cannot be legitimately dissociated from it. Therefore, military members – through socialization, formal education, and training – structure and shape their professional identity around these values. It is important to emphasize here the distinctiveness of identity shaped by Canadian society. Because military members are coming from and returning to the society, we are assuming that their identity is in concordance with "the parent community,"[21] and that they have embraced and integrated cultural values that are defended by Canadian society. However, the integration of cultural values through perception and interpretation can differ from one person to another, as each individual's process of *making meaning* is in part unique to his/her own subjective experiences. Therefore, it is assumed that an organization such as the military is confident that all of its members understand and interpret values the same way that are (supposedly) commonly accepted as part of a democratic society.

The central importance of values is noticeably expressed in *Leadership in the Canadian Forces: Conceptual Foundations* as it reinforces the CF's adoption of value-based leadership. Such leadership provides "direction" and helps members to find their way through challenging situations to arrive at appropriate decisions. "A key proposition of values-based leadership is that the guidance provided by the CF values is the ultimate recourse and compass when explicit direction from superiors is lacking, when dealing with ambiguous situations, or when operating under competing demands and pressures."[22] The above statement structures the CF around the concept of values as the primary vehicle through which military members' professional identity is developed and their process of *making meaning* and/or moral reasoning is informed.

Second Level: Personal Values

While values at a personal level are as important – and maybe more so – than at the professional level, it is, however, difficult to impose them on others or to simply express them as if they were part of a larger system, in this case the military organization. In most circumstances, military members' personal values are congruent with the profession in which they serve. In those circumstances, there are no conflicts or competition between values. In other situations, military members have strong personal values that come into conflict or are difficult to convey while serving in the profession of arms. Furthermore, some may not even be aware that certain personal values (e.g., discrimination against minorities) interfere with their profession. In these two latter cases, personal values also shape the professional military members' identity and inform their way of *making meaning* of their work and surroundings. That is, tension between personal and professional values results in a process of evaluation in which the individual must decide what is "right"; this evaluative process (i.e. moral reasoning) always occurs within their cultural context or milieu.

One unquestionably important example of personal values in today's forum, especially in the context of the "war against terrorism,"[23] is the question of religion. This same example will later guide the discussion of Kegan's identity development framework. The question that is asked here is not one of "for or against that war" but one that addresses the reality *of* and *in* war. In other words, how does religion as a personal value affect that reality? While this is a difficult and complex issue, this chapter emphasizes only one of its aspects: that religion – consciously or subconsciously – continues, to a certain extent, to inform the reality *of* and *in* war by defending a dualistic viewpoint that classifies behaviours. This dualistic way of making sense of the world presents constructs such as: good and bad; honest and dishonest; acceptable and unacceptable; loyal and disloyal; right and wrong. The argument here is that there is a cultural foundation (i.e., religious values and beliefs) on which other values (i.e., professional and personal) are based and, to a large degree, continue to be maintained or adapted. Thinkers such as Jean-Claude Guillebaud say that, "our democratic values (e.g., equal rights) are linked to our Jewish and Christian heritage, secularized during the Enlightenment."[24] Further, Canadian philosopher Charles Taylor says that, "…our sense of where we are is crucially defined in part by a story of how we got here. In that sense, there is an inescapable (though often negative) God-reference in the very nature of our secular age."[25]

We argue, along with sociologist Roger O'Toole, that our cultural values are influenced and remained deeply rooted within a Judeo-Christian philosophy:

> The persistently *Christian* [emphasis original] character of Canada, in a broad sense, is an important legacy of this past century and a frequently underestimated fact of considerable sociological interest. Despite the impact of secularization, an apparent crisis of religious commitment and a rapidly expanding non-European presence, Canada remains decidedly Christian…Moreover, a recent expansion of numbers of those embracing religions other than Christianity together with an increase in those professing no religion has not altered this state of affairs to any significant degree. Canada remains a society where Christian traditions with historical roots in Britain and Western Europe dominate the demography of religious identity from Newfoundland to British Columbia.[26]

This continuous influence has been expressed in debates in the province of Quebec, in which people have continued to challenge the meaning of religious accommodations with arguments that take into consideration beliefs, values and traditions that were contested during the Quiet Revolution.[27] It seems that a situation such as a debate on religious accommodation revives old connections to a religious past and brings to light the complexity of cultural legacy and background. The matter becomes more complicated when members of the CF, as an example, encounter similar situations in an unfamiliar environment (e.g. Afghanistan) where people strongly adhere to other religious beliefs, traditions, and ways of living.

As mentioned above, values at the professional and personal levels shape and inform CF members, and are the underlying condition for an individual's decision-making and, more importantly, for their conduct in familiar and unfamiliar environments. Given the underlying link between culturally-instilled value(s) systems, such as religion, to an individual's moral reasoning and behaviour, it is evident that those in different cultural settings with different instilled value(s) systems will understand the (social) world and what is "right" or "wrong" in different ways. Thus, the first step in bridging the cultural gap is to examine *how* value(s) systems (i.e. religion) come to shape and inform identity development.

KEGAN'S MODEL OF IDENTITY DEVELOPMENT

This chapter suggests that CF members need to understand, interpret, and integrate the values to which they adhere and defend in and outside of Canada, in order to develop the capacity to understand, interpret, and integrate the values that *other(s)* defend and thus make sense of their meanings. *Making meaning* – understood as "a meaning-constitutive activity"[28] – of one's interpretation of one's own values and beliefs and those of *other(s)* stresses that the notion of values (*own* and *other(s)*) influences the way military members, at all levels, see themselves as CF professionals. In other words, *making meaning* shapes CF members' professional identity, which directly influences what they understand to be the "right" action and thus how they conduct themselves.

In that context, the following questions should be asked: *how* do CF members construct their sense of identity as military members in a complex and ever-changing environment that requires an openness to unfamiliar realities, customs, traditions, and values? *How* do CF members understand themselves as military professionals in cultures where the expressions of such values and traditions are contrary – or seem contrary – to dominant Western realities? The *how* is the centre of gravity that sustains the fundamental and important process of developing a more complex level of *making meaning* among Canadian military members' *vis-à-vis the other(s)*.

According to Robert Kegan, this capacity of *making meaning* is associated with the structure of someone's identity development.[29] The concept of identity is understood as a process, characterized as an "evolution of meaning."[30] Kegan suggests that people evolve towards an increasingly complex way of seeing themselves and the world. The degree of complexity refers to a change in *meaning-making*, where internal and external experiences become progressively more integrated in articulating a broader view of oneself, others, and the world. This evolutionary process does not happen in a vacuum; it implies a change in a way of thinking and requires the ability to analyze multiple aspects of a particular situation objectively. To a certain extent, this complex way of thinking means a deeper consideration of what is required when a particular situation or context – such as an unfamiliar cultural setting or event – is perceived. In other words, it is *how* a person organizes the meaning of such a situation. The focal point here is the individual experiencing the *meaning-making* to the point that he/she becomes the *act* of the *meaning-making*. Kegan says that: "What a human organism organizes is meaning. Thus it is not that a person makes meaning, as much as that the activity of being a person is the activity of meaning-making."[31]

At this point, it is important to identify which level of identity development is required for military members in the context of CQ. At which level, for example, would a military member be able to understand and integrate what constitutes behaviours that are culturally embedded in the environment of *other(s)*? In a practical sense, identity is used as a "filter" for constructing and organizing new understandings of unfamiliar cultural settings. This "filter" enables a person to be effective regardless of the cultural environment he/she is in.[32]

Robert Kegan has identified 5 stages for explaining this evolutionary process, which regards identity development as a personal construct that tends to close the gap between a situation (seen as an object) and a person (seen as a subject). In most cases, a person tends to see a situation as organized in a *vis-à-vis* or opposite and separate manner from himself/herself as the observer; this type of relationship will determine his/her grounds for constructing the meaning of this situation. In contrast, Kegan argues that the relationship between a subject and an object is not *vis-à-vis*, but the subject and the object are invested in the same movement when trying to make sense of a particular situation. In other words, by experiencing a situation, the person is an integrated part of the very situation that he/she is observing.[33]

The features of Kegan's stages focus on the aspect of the *self* that evolves towards an understanding of the cultural systems and groups that shaped it and of which it is a part. To be more precise, Kegan stresses the fact that a person as a subject is already invested within situations (e.g., social groups, or working environments such as the military) that mark his/her identity to the extent to which he/she expresses it – consciously or subconsciously – through his/her behaviours (see Table 5-1). In order to illustrate the focus of these stages, the case of religious values, beliefs or associations will be used as an example. This case tries to show how, in the context of Afghanistan where religious beliefs, values, customs and traditions are the cultural norm and thus embedded in daily activities, military members understand, perceive, integrate and identify themselves in relation to their (unfamiliar cultural) surroundings.

STAGE	MEANING-MAKING	IMPACT ON CQ
1. Single Perspective	Focus on only one aspect of a situation, allowing for only one possible way of seeing things.	In this stage, it is difficult to recognize that other individuals have their own purpose and/or way of seeing things. **No CQ**.
2. One Perspective at One Time	Aware of more than one's own perspective; but only capable of dealing with one distinct view at a time: one's own and the others'.	The individual is receptive to the existence of different points of view, but treats them separately, i.e. attempt at gathering knowledge of the other. **No CQ**
3. Opening to Reciprocal Perspectives	Aware of shared values and experiences that take precedence over individual interests. Individual's way of seeing things is supplemented with that of the other.	An attempt to construct and share an alternative that represents an internal interest of the other. Meaning behind the other's perspective in context is comprehensible. **Minimal CQ.**
4. Internalization of the Other	Self-identification with the other; akin to empathy. The individual is engaged culturally in a dialogue with the other.	The individual is able to empathize with the other's perspective; to see the world through the eyes of the other, to an extent. Able to recognize core elements of the other's cultural world view. **Moderate CQ.**
5. Universal Perspective	The individual's perspective is understood in the context of a global experience of humanity.	**Advanced CQ.**

TABLE 5-1: KEGAN'S IDENTITY FRAMEWORK AND CULTURAL INTELLIGENCE

Before addressing this framework, it should be emphasized that Stage 5 (and to a certain extent Stage 4) of Kegan's framework is perhaps above and beyond what the CF organization can achieve at this point. However, this chapter argues that these two stages are required at the higher rank because of the higher CQ requirement at that level. Moreover, in terms of building the CQ capacity of CF members in general, there is a minimum requirement to operate at Stage 3.

Stage One: Single Perspective

The difference between the ways women are "viewed" in Canada and Afghanistan is an example of the first stage. According to Canadian cultural values, men and women are to be treated equally and any deviation from that "right" or cultural value is highly criticized. Therefore, a Canadian soldier's view of gender relations in the Islamic world, informed by Western cultural standards, might be that women are subordinated to men. This one-sided perspective could taint the CF member's perception of what the Islamic world or Islamic faith represent. In other words, at Kegan's first stage of development, there is no attempt to go beyond this first perceived "reality" of the situation.

Stage Two: One Perspective at a Time

In the second stage, there is an attempt to recognize other aspects of an issue or situation. In the example above, the treatment of women negatively impacts the perception of what the Islamic world or Islamic faith represents because of the evaluation according to Western standards. While that issue still impacts an individual's perception, at Stage 2 there is openness to other perspectives that might be more accurate of the Islamic world or Islamic faith. The custom of wearing the traditional *hijab* or *burkah*, as an example, *could* be perceived as a sign of modesty rather than submission. At this stage, an individual is receptive to these perspectives but, as previously mentioned, they are treated separately.

Stage Three: Opening to Reciprocal Perspectives

In the third stage, the broader context of the Islamic world or Islamic faith is envisaged and differences in customs (e.g., *hijab*, daily prayers, Ramadan) are comprehensible and even acceptable in that particular (cultural) context. Furthermore, in the context of being a Canadian soldier, there is an attempt to transmute this role from one of being merely a combatant committed to a "fight against terrorism" to one characterized by adaptive behaviour directed towards the restoration of a better way of living.

Stage Four: Internalization of the Other

At Stage 4, the individual has internalized different perspectives (e.g., history, politics) and aspects (e.g., practices, movements) of the Islamic world and Islamic faith. There is a (*self*)-identification – akin to "empathy" – that is internal to the person in a way that denotes a capacity to transition towards a view of the Islamic faith as it is completely embedded within the Islamic cultural world view. In other words, to more fully comprehend the cultural values and practices of the Islamic faith as it relates to the context of the Islamic world more generally. This change of position from one of exclusion (e.g. *either/or*) to one of inclusion (e.g. *oneself and other*) indicates that a person experiences and is engaged culturally in a dialogue with the *other(s)*. In other words, the individual seeks to experience multiple perspectives or facets of a situation (e.g. religious experience in the Islamic world) by not opposing them but by relating with them. Moreover, these perspectives and facets of the *other(s)*' experience may also represent perspectives and facets of one's own similar experiences (e.g. debate on religious accommodation in Quebec). Therefore, there is often congruent proximity between experiences.[34]

Stage Five: Universal Perspective

Stage 5 represents a higher order in the development of a person's identity. The higher order recognizes a plurality of systems (e.g. ways of thinking and interpretation); cultures (e.g. Canadians, Americans, and Muslims); organized religions (e.g. Catholicism, Protestantism, and Islam); and groups (e.g. Christian and Islamic fundamentalists) that are interrelated and conceptualized as "universal". The "universal" designation is understood in the context of a global experience of humanity and as shared principle to defend. The "universal" designation becomes as Charles Taylor says: a "shared moral space" that links one human to another.[35] In that perspective, a Canadian soldier *makes meaning* of what is *different* by reaching out – beyond differences – to the human experience of the *other(s)*. Religion is no longer viewed as an issue of controversy unto itself, but is an element of a larger cultural framework or system that defines a group or an organization. A given culture, as an example, is not only understood as a contextual phenomenon shaped by customs and traditions, but as an observable way of life informed by perceptual, linguistic, and practical human *experiences*.[36]

As a mentioned above, Stage 5 of Kegan's framework is, practically speaking, above and beyond what the CF organization can achieve at this point. In terms of building their capacity for CQ, however, CF members need to operate at a minimum of Stage 3. Members must be able to internalize the world view of the *other(s)* to the point where they can at least understand their values and practices as "right" within their cultural context. It must be noted, however, that developing an understanding of how another culture's values make sense in their own environment must be accomplished without falling into the trap of *cultural relativism*. That is, having the capacity to understand how the other's cultural perspective is "right" for them should not cause an individual to abandon their own cultural understanding of "right" and "wrong". Instead, a CF member who has achieved Stage 3 of Kegan's framework is able to understand the other's values without having to judge them as "right" or "wrong" because they are the basis of the way of life of the people in that cultural setting. Having achieved this understanding, the CF member must strike a balance between "doing the right thing" by Western standards, and "doing what is right for the *other(s)*" according to the cultural context in which they are operating.

DISCUSSION

The objective of this chapter was to articulate the cognitive dimension of CQ as it is relates to Kegan's identity development model. It was demonstrated that the concept of identity is a process that evolves from one stage to the next, and that it corroborates the construct of a broader internalization of the *(one)self* and of the *other(s)*. This model stresses the necessity for the individual to recognize the *self's process of meaning-making* when dealing with complex issues.

Canadian military missions like Rwanda, Bosnia, and now Afghanistan confirm the need for military members to become culturally adept or at least competent when encountering what this chapter has called the "unfamiliar." It is clear that the first test for Canadian military members is the challenge of becoming cognizant of their own cultural values and assumptions. At this point, it is not a question of conflicting or non-conflicting values. It is an issue of knowing (i.e. content), understanding (i.e. learning) and internalizing (i.e. identification with) values that concern the *raison d'être* of an embraced profession (e.g. the profession of arms), and cultural values (e.g., family, societal, religious and educational values) that have formed an individual's identity. When concerned with a desired outcome founded on a values-based approach, military members have a direct influence on the way a mission is accomplished. Again, it is not a problem of evaluating who is "right" and who is "wrong"; the challenge faced by Canadian soldiers is one that addresses the issue of "*who* is to judge what makes something right and wrong?" There is an indisputable expectation for Canadians to do "what is right" and at the same time embrace a seemingly "relativist viewpoint" that might contradict what they define to be "doing right." The ability to strike a balance in this situation is the result of successfully applying CQ.

Kegan's framework of identity development challenges the so-called practice of "cultural awareness" by inviting military members to transcend a particular knowledge that focuses of *what is seen* and *perceived* (i.e. learned knowledge) and to be engaged within a discourse that requires a knowledge that is *culturally integrated* (i.e. developmental knowledge) with their professional development. Such developmental knowledge necessitates combined prerequisites: training and education. The military member needs to be trained (e.g., technical knowledge) and educated (e.g., imperative knowledge) in order to function in ambiguous situations that could create "cultural tensions" because of the nature of the behaviours and practices they are required to embrace in order to connect with the other(s). It is therefore important to recognize the necessity for military members to train and educate their way of *thinking* or *meaning-making*. As an example, the case study presented at Appendix A illustrates the way in which a CF padre's recognition of the cultural values of the *other* was used to effectively influence Pashtun perceptions of Canadians as the *other*.

The crucial issue of expanding military members' *way of thinking* or *meaning-making* depends greatly on members being involved in their own evolving process of identity development. To be precise, both necessities (way of thinking and identity development) cannot be separated. While this is a particular challenge, it is not impossible because professional development (through training and education) at all levels is already inviting military members to engage themselves in a more holistic approach that incorporates different aspects of what is required when confronted in a more complex and ever-changing world.

The PDF, which is presented and discussed in greater detail in Chapter 6, is a good example of how a military member is invited to integrate the identity development process into his/her professional development. From the junior to the senior level, there is an evolving process that takes an individual from the more rudimentary to the more complex understanding of what is necessary for members of the profession of arms. The identity development process outlined in this chapter is intertwined like a spiral with capacities such as cognitive capacities (e.g. abstract knowledge), change capacities (e.g. paradigm shifting) and professional ideology (e.g. being steward of the profession).

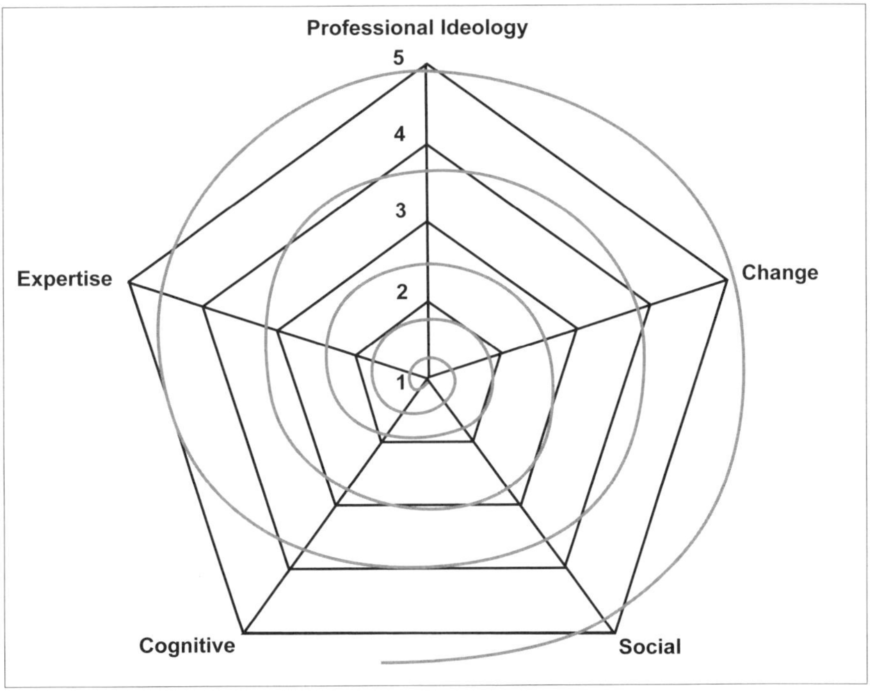

FIGURE 5-1 : KEGAN'S IDENTITY DEVELOPMENT SITUATED ACROSS THE PROFESSIONAL DEVELOPMENT FRAMEWORK CAPACITIES

Figure 5-1 illustrates a suggested model for conceptualizing Kegan's identity development process situated in the context of the PDF. Notably, each of the five PDF capacities has equal weight within each of Kegan's five stages. That is, identity development at any one stage cuts across all PDF capacities equally. Further, as identity development progresses, each stage of development is embedded in the next; none of the stages are mutually exclusive. Finally, Kegan's Stage 5, the universal perspective, is situated outside, signifying a global experience.

SUMMARY

Recognizing that one's process of *making meaning* and therefore one's understanding of "right" and "wrong" is informed by, and developed through, the values and practices of the culture or professional organization with which one identifies is crucial to developing a capacity for cultural competence or CQ. For the CF, recognizing such a process of *making meaning* is more than allowing military members to inform (and modify) their behaviours with a new cultural competence.[37] It also requires openness to a *different* knowledge characterized by a *malleable or adaptive structure*, and as a new and unfamiliar association that serves as the foundation for building common grounds with the *other(s)*.

NOTES

1 In *Cultural Intelligence: People Skills for Global Business*, David C. Thomas and Kerr Inkson present a framework of developmental stages of CQ. While this chapter does not adopt this framework it does, however, recognize some similarities with Kegan's identity development process. (San Francisco: Berrett-Koehler Publishers, Inc., 2003).

2 It is interesting to see that in 2007, the Center for Strategic and International Studies in Washington produced a report regarding the place of religion in conflicts. See, *Mixed Blessings: US Government Engagement with Religion in Conflict-Prone Settings*, (August 2007).

3 Robert Kegan, *The Evolving Self: Problem and Process in Human Development*, (Cambridge: Harvard University Press, 1982), 103.

4 Jean Piaget (1896-1980). Genevan developmental psychologist, particularly well-known for his stage theory of cognitive development.

5 Lawrence Kohlberg (1927-1987), American psychologist who pioneered the study of moral development in the late 1950s.

6 Michel Foucault, "The Ethics of the Concern of the Self as a Practice of Freedom," in Paul Rabinow, ed., *Michel Foucault: Ethics, Subjectivity and Truth*, Volume 1 (New York: The New Press, 1997), 286.

7 Christopher Earley and Randell S. Peterson, "The Elusive Cultural Chameleon: Cultural Intelligence as a New Approach to Intercultural Training for the Global Manager," *Academy of Management Learning and Education*, 31 (1, 2004), 106

8 Ibid.

9 Ibid.

10 Ibid., 107.

11 The concept of oneself is used here in reference to the French terminology "soi-même." This chapter does not address the lengthy discussion about the "constitution" of oneself. For more discussion on the subject, see Anthony Elliott, *Concepts of the Self*, (Cambridge, UK: Polity Press, 2001).

12 In "Military Cross-Cultural Competence: Core Concepts and Individual Development", anthropologist Brian R. Selmeski goes further and says that: "It is also critical to acknowledge [...] that culture is not a social group, material object, activity or officially articulated statement." Occasional Paper Series – Number 1, Centre for Security, Armed Forces & Society, (Kingston, ON: Royal Military College of Canada, 2007).

13 The *other* could also be understood as the *other monotheism* as mentioned by the psychoanalyst Julia Kristeva: "The one and the other are two aspects, semantic and logical of the imposition of a strategy of identity, which is, in all rigor, that of monotheism." *Powers of Horror: An Essay on Abjection*, trans. L.S. Roudiez, (New York: Columbia University Press, 1982). Cited in Mark C. Taylor, *Altirity*, (Chicago/London: The University of Chicago Press, 1987), 172.

14 Maurice Bellet, "N'importe qui [...] qui me concerne," *Le Point critique* (Paris: Desclée de Brouwer, 1970), 216.

15 "An emphasis on values orientation and understanding others through their related beliefs and practices underlies much of current work on intercultural training and management education." Christopher Earley and Randell S. Peterson, "The Elusive Cultural Chameleon," 101.

16 *Oxford Dictionary of Philosophy*, (Oxford: Oxford University Press. 1994), 390.

17 Michel Maffesoli, "Towards a Postmodern Ethic of the Aesthetic," *The Future of Values: 21st Century Talks.*, Jérôme Bindé, (ed.), (Paris: UNESCO Publishings, 2004), 69.

18 Mohammed Arkoun, "For a Subversive Genesis of Values," in *The Future of Values: 21st Century Talks.* Jérôme Bindé (ed.), Paris: UNESCO Publishings, 2004). 50.

19 Canada, *Duty With Honour: The Profession of Arms in Canada*, (Kingston, ON: Canadian Leadership Institute Leadership, 2003); and Canada, *Leadership in the Canadian Forces: Conceptual Foundations*, Kingston, ON: Canadian Forces Leadership Institute, 2005).

20 Canada, *Duty With Honour*, 26-27.

21 Ibid. 28.

22 Canada, *Leadership in the Canadian Forces: Conceptual Foundations*, 16

23 Someone may also argue that the language (e.g., evil, crusade) used in certain discourses has a religious connotation "Language that hints of superiority is especially problematic since

the current war on terrorism targets the Islamic world, and the rhetoric is essentially that of a crusade agains Islam." Timothy L. Challans, *Awakening Warrior: Revolution in the Ethics of Warfare*, (New York: State University of New York Press, 2007), 13.

24 « Cela signifie que la plupart des valeurs démocratiques (égalité...) ont partie liée avec l'héritage juif et chrétien, laïcisé à l'époque des lumières. » Jean-Claude Guillebaud, *Le gout de l'avenir*, (France: Éditions du Seuil, 2003), 323.

25 Charles Taylor, *A Secular Age*, (Cambridge: The Belknap Press of Harvard University Press, 2007), 29.

26 Roger O'Toole, "Canadian Religion: Heritage and Project," in D. Lyon and M. Van Die, (eds.) *Rethinking Church, State, and Modernity: Canada Between Europe and America,* (Toronto: University of Toronto Press, 2000), 45, cited in Joanne Benham Rennick, *Religion in the Ranks: Religion and Military Culture in the 21st Century,* Doctoral thesis (close to completion), (Waterloo: ON, 2008), 58. Personal copy.

27 In February 2007, for example, the Premier of Québec, M. Jean Charest, established a consultation commission, to be headed by Sociologist Gérard Bouchard and Philosopher Charles M. Taylor, on accommodation practices related to cultural and religious differences, following extensive debate within Québec on related issues. The commission report, along with 37 recommendations, was released in May 2008. <http://www.accommodements.qc.ca/commission/>, accessed Jun 2008.

28 Robert Kegan, *The Evolving Self*, 42.

29 For a study using Kegan's framework, see G. B. Forsythe, S. Snook, P. Lewis and P. Bartone, "Making sense of Officership: developing a professional identity for 21st century army officers," in L. J. Matthew, (ed.), *The Future of the Army Profession*, (New York: McGraw-Hill Primis Custom Publishing, 2002), 357-378.

30 Robert Kegan, *The Evolving Self*, 15.

31 Ibid., 11.

32 P. Christopher Earley and Soon Ang, *Cultural Intelligence: Individual Interactions across Cultures*, (Stanford, California: Stanford University Press, 2003), 11; Christopher Earley and Randell S. Peterson, "The elusive cultural chameleon," 106.

33 Robert Kegan, *In Over Our Heads: The Mental Demands of Modern Life*, (Cambridge: Harvard Press, 1994), 32.

34 In our point of view, this proximity could also be called "mindfulness." According to David Thomas, "mindfulness is fundamentally a heightened awareness of and enhanced attention to current experience or current reality." David Thomas, "Domain and Development of Cultural Intelligence: The Importance of Mindfulness," in *Group & Organization Management*, 31 (1 February 2006), 84. Also see discussion of mindfulness presented in Chapter 2 of this volume.

35 Charles Taylor, *Les sources du moi. La formation de l'identité moderne*, (Paris, Éditions du Seuil, 1998), cited in Guillebaud, 246.

36 Gary Gutting (Ed.), *The Cambridge Companion to Michel Foucault* (2nd Edition), (New York, NY: Cambridge University Press, 2005), 79.

37 It is important to mention that the experience of *oneself* and the *other(s)* is a reciprocal one. The *other(s)* has to become also the *self*.

RECOMMENDED READING

Bindé, Jérôme (ed.). *The Future of Values: 21st Century Talks*. Paris: UNESCO Publishings, 2004.

Earley, Christopher and Randell S. Peterson. "The Elusive Cultural Chameleon: Cultural Intelligence as a New Approach to Intercultural Training for the Global Manager," *Academy of Management Learning and Education*, 31 (1, 2004).

Kegan, Robert. *The Evolving Self: Problem and Process in Human Development*. Cambridge: Harvard University Press, 1982.

Moore, S.K. and Suleyman Demiray, "The Canadian Forces Chaplain Branch: Modelling Interfaith Cooperation and Pluralism in Afghanistan," *Ecumenism*, 165 (2007), 5-8.

Taylor, Charles, *Les sources du moi. La formation de l'identité moderne*. Paris, Éditions du Seuil, 1998.

Taylor, Charles, *A Secular Age*. Cambridge: The Belknap Press of Harvard University Press, 2007.

CHAPTER 6

DEVELOPING CULTURAL INTELLIGENCE

Dr. Robert W. Walker

The preceding chapters generated an enhanced understanding of Canadian society with its diversity and cultures, the CF with its culture of professionalism, and the CQ competencies required of CF members. This chapter addresses the challenge of developing that requisite CQ and emphasizes that the subject matter and values systems relevant to cultures, as a social phenomenon, need to be well learned, mastered and inculcated through a variety of learning processes and andragogical (adult-learning) PD strategies.[1] Each learning methodology must be more than a standard pedagogical (child-focused), linear-learning initiative involving passive student-receipt of teacher-centred presentations of dollops of cultural information.

The collective focus of this chapter is: first, to identify the urgency for the PD of CQ as applicable to various institutions; second, to identify those requisite leader elements and attributes that can generate strong CQ; third, to identify the appropriate learning strategies and methodologies with the highest potential to generate the requisites for strong CQ; and fourth, to explore specific CQ PD options for CF leaders and members.

INSTITUTIONAL EFFORTS AT CULTURAL CHANGE

Military organizations globally have been stereotyped, in the worst-case scenario, as hierarchical, hyper-conservative and isolated organizations with well-disciplined followers of autocratic, arrogant and possessive superiors who oversee the unique profession of arms, see themselves as the only profession that understands effective leadership, and resist influence and interference in warfighting and national security from politicians, bureaucrats, academics, think-tank researchers/advisors, and citizens in general. Navies, armies, air forces, as well as other uniformed organizations, e.g., law enforcement/policing institutions and correctional services, are held up, often mistakenly, as common examples of such institutions.

During the last five decades of the 20th century, military institutions and police organizations experienced major global, national, societal and external pressure to significantly strengthen internal accountability and to restructure in ways that would immediately change, and continue to change, stereotypes of isolated institutions and autocratic leadership. In response to this pressure, the CF, under General Rick Hillier, CDS from 2005-2008, moved through a Forces-wide transformation aimed at greater mission success. A 2009 CDA Press publication describes in detail the events, successes and challenges of the first years of its implementation.[2] Meanwhile in 2007, the Royal Canadian Mounted Police (RCMP), for the first time in its 134-year history, witnessed the obligatory resignation of its uniformed police Commissioner and his replacement with a governmental civilian Commissioner, following a series of questionable law-enforcement decisions and events, including an increase in the number of deaths of on-duty RCMP officers.

Before proceeding further, it is appropriate to identify the organizational circumstances in which culturally sophisticated leaders will need to function. Several Canadian historical examples of military and law enforcement organizations (that cover 50-year periods and could be expected to have had traces of the worst-case institutional stereotypes described above) represent cases of established, conservative, uniformed organizations that attempted, and still are attempting, with varying degrees of success, to undergo institutional and cultural change in response to external pressures and developments.

These Canadian examples, along with one US Army initiative, address the circumstances in which effective leaders need to possess and utilize the requisite leader elements and cultural sophistication, and to recognize and engineer the necessary cultural transformation that would be part of the greater institutional transformation. One might hold CQ in mind as the key variable for assessing and determining the level of success or failure of the cultural transformation in the following three examples, and might also speculate about any CQ PD that, had it been implemented in advance, could have improved the outcomes.

Canadian Forces Unification

In the 1960s, the serving government decided that the Royal Canadian Navy, Canadian Army, and Royal Canadian Air Force would amalgamate into a single military force: the Canadian Armed Forces. The implications of this reorganization were extensive; in particular, significant challenges stemmed from a dearth of insight and understanding of the impact that the firmly entrenched cultures of the three services would have on integration processes. This process, still ongoing and relevant to today's imperative of joint and integrated operations, is captured in a case study, *The Canadian Forces Unification: A Half-Century Clash of Cultures*, in Appendix B.

The Decade of Darkness

The 1990s were the tail-end segment of the "unification" years, often referred to by senior leadership as the decade of darkness. It was during this decade of darkness that the CF,

> ...particularly the officer corps ... had imploded and found themselves at the lowest ebb of their history. They had lost the confidence and trust of the government and the Canadian people they served.[3]

Colonel Bernd Horn and Dr. Bill Bentley of the Canadian Forces Leadership Institute (CFLI) described the CF as a conservative and anti-intellectual institution comfortable since 1949 with the perspective that the military knew best and could work without political or civilian interference.[4] John English also wrote extensively on this decline of Canadian military professionalism that continued through the decade of darkness.[5]

Synonymous with this internal turmoil, the major post-Cold War restructuring of international relationships took place along with regional pop-up wars and expanding terrorism. The CF was in an imposed transformation of stream-lining and down-sizing, followed by the "9/11" of 2001. Dynamic global warfare has evolved into circumstances of coalition warfighting and multinational co-operation in support of failing states. With these coalitions and multinational initiatives came the urgent requirement for CQ sufficient to improve world circumstances.

Community-Based Policing

The military was not the only uniformed and conservative institution attempting to change its culture and acquire CQ. For Canadian policing, the transformation attempted over the past 35 years has been an evolution from a virtually exclusive and somewhat elitist "professional police" model with a focus on criminal codes, responsive and reactive law enforcement, and gate-keeping for the justice system, and only a token "public relations" office, toward a "community policing" model that reflects a commitment to broad public safety, involves citizens in police priorities, planning, applications and relations, consults with the public about quality of life issues, and preserves democratic rights and values.[6]

Self-governing professions like law enforcement can be seen as privileged in that for generations, it has been left to them to enforce codes of ethics developed for their specialty of expertise – medicine, law (lawyers), law (policing), religion, and the profession of arms. For police organizations, the mission, vision, pressures, values, beliefs, desired results, criteria for success, etc., all differ in accordance with the perceived purposes of that profession by diverse components of society. The authority of conventional, "professional" police departments derives from the law and its enforcement, and the predominant influence on that authority and its priorities are those crimes already committed and that demand the most public attention. This is done through judicial/legalistic laws, regulations and punishments, e.g., criminal codes. Police are the experts in fighting crime, although they "diversify" to ensure neighbourhood peace and tranquility, or to regulate traffic flow or major-event crowd control. Social and community issues were seldom considered police responsibilities unless law and order were threatened.

The community-based policing model engages citizens and residents, or their representatives, in police activities that have the most direct impact within their district, thus reflecting community demographics, aspirations, concerns, fears and general goals. Police policy is developed through police-citizen liaisons in focus groups, "town halls", public relations, "community nights" and victim assistance initiatives. These two sets of "police culture" – professional law enforcement and community-oriented policing – reflect dichotomous concerns, actions and decisions and, as a consequence, planning procedures, regulations and policies.[7] Even the nomenclature of "police force" versus "police service", and the motto of policing, "To serve and protect", highlight the dichotomy in the philosophy of public safety. Naturally, these "two cultures" generate inter-departmental conflict, even light-hearted competition, among "true crime-fighters" versus "school patrollers" and "teddy bear cops". Leaders in policing need high CQ to manage these internal conflicts, while always recognizing the diversity of the cultures they serve and the needs, rights and freedoms of the citizenry.

The United States Army and Cross-Cultural Savvy

Interestingly, as a foreign example related to CQ, a US Army War College study identified a strong need for one particular strategic leader competency, which it labelled "*cross-cultural savvy*".[8] Described as far more than just the ability to work with non-American nations, militaries and citizens, this savvy was seen as fundamental to effective leadership and to comfortable and successful interactions with joint, international, interagency, or inter-organizational entities, including, according to War College guru Dr. Leonard Wong, the foreign warlords, foreign nations' government representatives and their citizens, as well as, not surprisingly, US congressional staffers, other government departments (OGD) and NGOs. Like concentric circles, cultural relations were seen, sequentially, as unique and different among diverse units and roles and relationships within a military force, between a military force and its ministry or government department, between a military and its citizenry with their diverse social values, and among militaries and their international military partners and governments. Dr. Wong's research was precipitated by the 2001 "9/11" terrorism attacks on the US, while the results and recommendations supported realignment and transformation of military leader PD to incorporate strategic leader competencies, including this cross-cultural savvy.

Similarly, CQ addresses an important dynamic within CF leadership and PD: the integration of cultural expertise, cultural sophistication, and cultural awareness with leader capacities to effect culture change and thus address both internal organizational dynamics and external influences at the national and international, military and government levels.

CULTURAL INTELLIGENCE, LEADERSHIP, AND PROFESSIONAL DEVELOPMENT

The changing nature of societies, the threatening security environment and rapid technological innovation necessitate an aggressive approach to military transformation and, therefore, to the continuous PD of military members. The CF initiative is a transformation focused on people – professional members with sophisticated leadership capacities, particularly in the spheres of cognitive, social, and change capacities, to master the new and expanding global challenges, as well as the internal transformation itself. It should be emphasized that, just as leadership and its complexities make PD of leadership so much of a challenge, so it is with CQ. Leadership is a layered challenge, from junior, analytical, interpersonal leader behaviours to senior, creative and abstract, inter-institutional, stewarding leader behaviours; and so it is with CQ, from junior members' less complicated cultural awareness and socially non-offensive and responsive behaviour, to complex cultural knowledge and socially responsible initiatives across cultures.

The CF, being hierarchic and bureaucratic in nature, and inseparable from the regulating role of the profession of arms, is able to achieve institutional effectiveness through the leading of its people, mainly at the tactical level, and the leading of the institution at the operational and strategic levels. Its current transformation initiatives reflect the complexity and volatility of these unique yet integrated roles, magnify the demands for tenacity, decisiveness and versatility, foretell the increasing challenges for professionalism in leadership, and underscore the pronounced need to enhance these evolving leader capacities through effective learning strategies and methods. This need for andragogical, and admittedly complex but effective, learning strategies is most evident in the types of complex learning content needing to be addressed across the higher leader levels.

In a 2007 CFLI report, Dr. Kimberly-Anne Ford and Karen D. Davis pulled together CQ, leadership and the CF PDF to explain the major challenges of teaching, learning, acquiring and developing CQ by drawing parallels to the same challenges that exist for mastering leadership. Ford and Davis emphasize:

> "In recent years, the Canadian Forces has developed leadership doctrine as well as a professional development framework comprised of a five-element cluster of CF leader capacities: expertise, cognitive capacities, social capacities, change capacities, and professional ideology. In meeting today's challenges, CF leaders need to be able to adapt to new cultural settings and to positively engage in the complexity of

multi-cultural interactions within a dynamic landscape of potential belligerents and allies. Thus, cultural awareness as well as the competencies to effectively develop and apply cultural knowledge is a critical contributor to mission success."[9]

The Professional Development Framework

Extensive research conducted on leadership and professionalism in the CF generated the five requisite military leader elements to comprise the PDF – *Expertise, Cognitive Capacities, Social Capacities, Change Capacities*, and *Professional Ideology*.[10] These five elements need to be incrementally and progressively developed throughout the careers of CF leaders at all levels. The PDF cross-tabulates these five elements with four leader levels – junior, intermediate, advanced and senior – and expands upon the foundations of leadership and professionalism. In addition, effective learning strategies are essential to the successful implementation of the PDF.

Brief descriptions of the leader elements and their attributes are provided at Table 6-1. A total of 16 attributes required of all CF leaders are nested within these five elements. Each attribute in turn would consist of a collated grouping of position-, level-, and role-specific competencies that would be identified case by case, in line with human resources systems needs, e.g., recruiting, performance assessment, promotion requirements, succession planning, etc. The 16 attributes represent *fundamentally necessary, but not necessarily sufficient*, attributes of CF leadership. These attributes also reflect the "chronology" of some of the elements: for example, Expertise evolves through a career from technical/specialty finesse to comprehensive military/organizational wherewithal, and on to sophistication with strategic and institutional leader roles and responsibilities. In contrast, Professional Ideology needs to be "front-end loaded" as new CF members are introduced abruptly to military norms and expectations of positive behaviour.

The inter-relationship of the five leader elements is best depicted as an assembly of joined puzzle pieces, i.e., a schematic that would visually represent the interconnectedness and interdependency of the leader elements that, only collectively,would make effective leadership possible. Figure 6-1 reflects this inter-relationship. While Table 6-1 describes the five requisite leader elements, and their 16 attributes, i.e., the leader framework, needed to generate CF effectiveness, Figure 6-1 depicts the interconnectedness, interdependency, complexity and interrelatedness of the puzzle pieces of leadership. The remaining challenge is to advance leadership doctrine as applicable to the CF through this framework of requisite leader elements being arrayed against a continuum of increasingly greater leader responsibilities and roles, i.e., where "the rubber hits the road".

A FRAME-WORK OF 5 LEADER ELEMENTS	16 ATTRIBUTES (IN BOLD) WITHIN 5 ELEMENTS ACROSS THE LEADER CONTINUUM The focus, scope, magnitude of Competencies for responsibilities related to the leader attributes will vary with rank, leader level, position, etc., and usually increase with time in CF, rank, seniority and credibility.
EXPERTISE	Expertise consists of **Specialist** (Military Occupation Classification) **and Technical** (clusters, e.g., combat arms, sea trades, aircrew) proficiencies, an understanding and development of the **Military and Organizational** environments, and the practice and eventual stewardship of the profession of arms, with the capacities to represent and transform the system through applications at the **Strategic and Institutional** levels.
COGNITIVE CAPACITIES	Cognitive Capacities consist of: a problem-solving, critical, **Analytic**, "left-brain" competence to think and rationalize with mental discipline in order to draw strong conclusions and make good decisions; plus an innovative, strategic, conceptually **Creative**, "right brain" capacity to find novel means, "outside the box" ends, and previously undiscovered solutions to issues and problems.
SOCIAL CAPACITIES	Social Capacities consist of a sincere and meaningful behavioural **Flexibility** to be all things to all people, with authenticity, combined with **Communications** skills that clarify understanding, resolve conflicts and bridge differences. These capacities are blended with **Interpersonal** proficiency of clarity and persuasiveness, **Team** relationships that create coordination, cohesion, trust and commitment, and **Partnering** capabilities for strategic relations building internally and especially externally.
CHANGE CAPACITIES	Change Capacities involve **Self**-development, with risk and achievement, to ensure self-efficacy, **Group**-directed capacities to ensure unit improvement and group transformation, all with an understanding of the qualities of a CF-wide **Learning Organization**, applications of a learning organization philosophy, and the capacity of strategic knowledge management.
PROFESSIONAL IDEOLOGY	Professional Ideology consists of an acute awareness of the unique, theory-based, discretionary body of knowledge at the core of the profession with an **Internalized Ethos** whose values and beliefs guide the application of that knowledge. The discretionary nature of military knowledge requires keen judgement in its use and involves **Moral Reasoning** in thinking and acting, shaped by the military ethos. Professional Ideology underpins a leader exemplar with **Credibility/Impact** who displays character, openness, assertiveness and extroversion that ensures the necessary effect by and from the leader.

TABLE 6-1: THE LEADER FRAMEWORK – FIVE ELEMENTS, SIXTEEN ATTRIBUTES

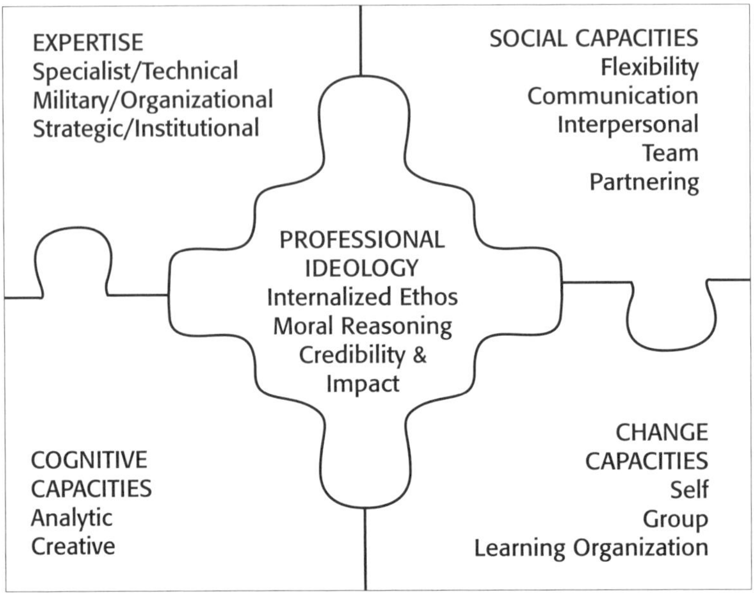

FIGURE 6-1: A LEADER FRAMEWORK – FIVE LEADER ELEMENTS, SIXTEEN LEADER ATTRIBUTES

The PDF at Figure 6-2, below, with an arbitrary "quartering" of the continuum of leader levels – junior, intermediate, advanced and senior levels – represents the full template for identifying, defining, and professionally enhancing the leader elements. Figure 6-2, however, even with a hypothetical progressive professional/leader development process stretched across its frame, is but a skeletal schematic of the leader elements and leader levels of the PDF.[11]

	Expertise	Cognitive Capacities	Social Capacities	Change Capacities	Professional Ideology
Senior	Strategic	Creative Abstract	Inter-Institutional	Paradigm Shifting	Stewardship
Advanced					
Intermediate					
Junior	Tactical	Analytical	Inter-Personal	Open	Internalize

FIGURE 6-2: THE PROFESSIONAL DEVELOPMENT FRAMEWORK – FOUR LEADER LEVELS, FIVE LEADER ELEMENTS

Professional ideology occupies a privileged position in the PDF. Significantly, professional ideology claims a unique, discretionary, theory-based body of military knowledge, and a military ethos that guides and adjudicates how that knowledge is used.[12] The theory-based knowledge consists of the General System of War and Conflict (introduced in Chapter 3) comprising policy, strategy, operational art and tactics. The military ethos consists of three components: beliefs and expectations about military service; fundamental Canadian values; and core military values, as articulated in *Duty with Honour: The Profession of Arms in Canada*.[13] Professional Ideology demands dedicated commitment over self-interest and effectiveness over efficiency.

When four of these PDF elements are shaped by the fifth, professional ideology, all five coalesce into a collection of interdependent puzzle pieces (as depicted in Figure 6-1) of the kind of leadership needed to achieve institutional effectiveness, in particular during institutional transformation and/or multicultural endeavours. So, beyond the challenge of effective PD of "just" expertise and military leader capacities, there is the complex PD relevant to CF leaders' professionalism and professional ideology. As well, it is evident that the more expansive leader levels – advanced and senior – are founded on the training, education and development taking place at the less expansive leader levels – junior and intermediate, and that the foci across leader levels reflect the transition in leadership from a predominantly "leading people" emphasis at junior and intermediate levels, to a predominantly "leading the institution" commitment at advanced and senior leader levels.

ANDRAGOGICAL LEARNING STRATEGIES

Why now? Why are such strategies required to learn CQ? The first three chapters of this book introduced the complexity of the challenges in learning CQ, integrating it with other human competencies and relating it to Canada's way of war. The successful integration of multiple leadership capacities requires integrated and complex learning strategies, i.e., andragogical, non-linear learning methodologies. The PDF with its five leader elements is a critical tool for achieving CF organizational effectiveness through the integration of the four enabling outcomes – *mission success, member well-being and commitment, internal integration*, and *external adaptability*[14] These outcomes are integrated with the conduct value of military ethos that infuses them with professional standards, professional ideology and professional effectiveness. Therefore, Institutional/CF Effectiveness = Organizational Effectiveness + Professional Effectiveness.

CQ is a phenomenon that permeates all quadrants of the CF effectiveness model and, like an umbrella, it covers all aspects of military service. It cannot be emphasized enough that CQ is a complex and multi-layered phenomenon that cannot be learned by passive, pedagogical, methodologies. Effective CQ is developed through learner engagement in active, andragogical learning strategies. CQ is symbolic of the challenges all military leaders face in a rapidly changing world. Achieving CF effectiveness requires a mastery CQ.

It is important to emphasize the symmetry of the PDF and the CQ. The PDF's five broad leadership elements – Expertise, Cognitive Capacities, Social Capacities, Change Capacities, with a pervasive Professional Ideology – closely parallel the dimensions of CQ – Knowledge, Cognition, Behavioural Skills, and Motivation (openness to learning/change) with a pervasive Mindfulness (self-awareness, empathy, perceptiveness). This PDF-CQ symmetry is indicative of the important relationships and integration of leadership, professionalism and CQ.

Fundamentally, then, the onset of the 21st century represents a post-Cold War period of profound complexity facing leaders as the CF institution reconstitutes itself; issues include the imperative of Forces-wide transformation, recognition of the uniqueness of an institution regulated by an evolving profession to meet the challenges of a changing society, an evolving security environment and a technological revolution. The challenges are significant and complex, just as the CQ mastery process, including androgogical learning strategies, is complex and demanding.

Comparative research on pedagogy and andragogy applicable and specific to Canadian military training and education is not plentiful, but it does exist. Three recent examples of research applicable to mastery of complex content and process through comprehensive and integrated learning strategies are described below: a focused Canadian naval officer PD study; Canadian public service initiatives to institute learning methodologies that reflect the culture of their organizations; and a Canadian military study on pedagogy and andragogy applied to the PD of senior military leaders.

Professional Development for Operational Naval Officers

A focused research initiative, the PhD dissertation of a naval commander, addressed the levels of expertise, and strategies for PD, of naval officers being trained to command ships.[15] Commander McCarthy-Senebald proposed six levels of expertise, then created and recommended a developmental process of situational-learning instructional strategies for such naval officer training. This research essentially uncovered the issues of recognizing strata for expertise in applied military settings, and pairing them with appropriate PD strategies from a pedagogical-to-andragogical continuum.[16]

Professional Development for Public Service Employees

The Canadian School of Public Service, as well as the Centre for Intercultural Learning in DFAIT, are institutions practicing andragogy, emphasizing the need for dynamic continuous learning as an essential aspect of the culture of their organizations. The School of Public Service, for example, is committed to a blended approach to learning:

> ...using a combination of classroom training, workplace learning, computer-assisted learning, conferences and special events, roundtables, seminars, as well as research and networking... [It] will explore the use of modern learning techniques such as action learning, communities of practice and coaching circles... [and] will also explore new delivery methods such as web casting, video-conferencing, business conferencing, e-learning and distant learning via satellite...[17]

The profile of an interculturally effective person developed by DFAIT includes numerous behavioural indicators, such as attempts "to enhance communication by avoiding any stereotypical presumptions about how local people would understand what is being said,"[18] that also represent learner-engaged strategies and goals that can be practiced in various settings.

Professional Development for Canadian Forces' Institutional Leaders

Research specific to andragogy and directly applicable and specific to CF PD of leadership and professionalism is also very rare, but it too exists. Commodore Jennifer Bennett studied effective PD strategies for senior CF leaders.[19] She contrasted the development of technical, thinking and interpersonal skills during the Cold War era, through pedagogical methodologies, with the andragogical learning strategies now needed to generate knowledgeable, adaptable, innovative, and culturally sophisticated 21st century CF leaders. Her research,

> ...confirmed that a more holistic, andragogical (student-centred) approach that includes effective assessment and feedback tools, mentoring, a CF leader framework, more frequent rotations through headquarters, flexible terms of service and increased opportunities for varied responsibilities that support leader development across a military career would enhance professional development of senior CF leaders.[20]

Bennett's work determined that adult learning styles that facilitate active learning as opposed to passive reception of instruction, resulted in more learning and better learning. Mature adults took on increasing responsibility and initiative for life's PD, evolving to more learner-centric approaches, e.g., e-learning, distance learning and self-directed learning.[21] For strategic leadership development, the most effective PD methods were interactive, focused more on process and less on content, and used case studies, role-playing, simulations, mentoring (relatively "new turf" in the CF)[22], coaching, syndicate discussions, 360° or multi-rater assessment feedback, and other relevant feedback, self-evaluation and reflection.[23] She presented this variety of methods as modern, innovative, flexible, timely, real-life and practical, and as andragogical methodologies that encouraged learners to increasingly assume responsibility for their own learning and, hence, their own effectiveness.

Bennett's military research identified the "Top 10" PD methodologies provided by CF respondents.[24] The more interactive methods ranked higher on the scale than the passive and self-directed methodologies, reflecting enthusiasm for hands-on/"minds-on", engaged forms of PD. Such findings were consistent with other research literature on adult learning,[25] particularly executive PD, that recommended "action learning" approaches for engaging the leader in finding real solutions to real-world problems, thus working from or extending beyond a foundation of previous experience and learning.

To summarize, Bennett and others emphasize that active, adult learning styles result in more, more substantial, faster, and more efficient learning. Recommendations are for:

- active learning, learner-centric learning and learner-centric approaches, e.g., e-learning, distance-learning and self-directed learning;
- strategic leadership development through a focus on process over content, using various methodologies – simulations, mentoring, coaching, 360°/multi-rater feedback, others;
- systematic comparisons of benefits and limitations of experience, education, training and self-development;
- confirmation of the "Top 10" learning strategies identified by CF military personnel;
- using additional research to address adult learning, particularly executive PD, that recommends "action learning" approaches for engaging leaders; and
- further research on, and exploration of, learning strategies appropriate to address the more complex challenges of strategic and institutional leaders.

LEARNING STRATEGIES FOR PROFESSIONAL DEVELOPMENT FRAMEWORK ELEMENTS

As explored above, complexities facing CF leaders include CF Transformation; a hierarchic and bureaucratic institution with a profession of arms; the changing nature of society; the 21st century security environment; rapid technological innovation; an evolving professional ideology; as well as a multi-dimensional model of CQ (see Figure 2-1) with complex and interrelated concepts. With such complexity to be mastered, the process of mastery, naturally, needs to be complex. The research consistently supports this perspective, but what does it all mean for actual implementation of learning strategies?

A synthesis of findings from the three research sources discussed in the previous section underscores the important differences between pedagogy and andragogy. Excellent material on leadership PD published by the Centre for Creative Leadership contains recommendations that have equal application to CQ and its PD.[26] PD of both leadership and of CQ generates learning circumstances where adults seek autonomy in their learning; prefer self-directedness; are goal-oriented; are problem-centred; need to know why they need to learn something; are practical as problem-solvers; have life experiences to apply; and are motivated differently, including opportunities to connect socially, meet external expectations, and derive enjoyment from intellectual stimulation. Such factors need to be considered in the design of learning strategies and methods beyond pedagogy.

To a substantial degree, then, this research collectively stresses that effective learning strategies, specifically for the five PDF leader elements, are necessary in order to generate a mastery of CQ that responds to the volatile multinational challenges. Strategies and methodologies generally need to be balanced between those appropriate for the more concrete elements (e.g. expertise) – where acquisition of fundamental skills and knowledge is emphasized predominantly through justifiable pedagogical, group-taught, teacher-centred, lecture approaches to information dissemination – and andragogical, student-centred, self-engaged, experiential learning methodologies appropriate for the less concrete and more abstract capacities (e.g. complex cognition, social/behavioural flexibility, the shaping of change, and professional ideology).

DEVELOPING CULTURAL INTELLIGENCE

Chapter 5 presents an excellent integration of Robert Kegan's five-stage identity development model[27] superimposed on a pentagonal depiction of the PDF's five leader elements (see Figure 5-1). The schematic depicts a spiralling outward of identity stages intertwined with PDF leader elements, e.g., change capacities as paradigm shifting, and professional ideology as evolving toward stewardship, while the text of the chapter explains CQ PD with respect to identity evolution and competency development.

Others have also attempted to describe strong CQ PD. In *Cultural Intelligence*, David Thomas and Kerr Inkson present their findings on increasingly successful functioning in multicultural environments.[28] They identify five developmental stages of CQ:

- reactivity to external stimuli by parochial individuals not recognizing cultural differences;
- recognition and awareness of the cultural mosaic, of cultural norms, and possession of a motivation to learn more about them;
- accommodation of other cultural norms, understanding of variations, and the absence of absolutes in cross-cultural perspectives;
- assimilation of diverse cultural norms into alternative behaviours and adjusting to diversity in situations without much effort; and
- proactivity in cultural behaviour based on recognition of changing cues unnoticed by others such that higher-CQ people sense changes in cultural context and act accordingly.

Thomas and Inkson determined that higher levels of CQ involved a cognitively complex perception of diverse environments, and the describing of people and events in terms of many different characteristics. They placed emphasis on exploration – trying out new behaviours while being mindful of the effects, learning from new interactions, and attempting to understand underlying thought processes and motivational aspects. These circumstances and challenges forcefully outlined the degree of commitment, energy and time required to elevate CQ to a level (Stage 5 in Thomas and Inkson) at which sophistication is broad, generic, and foundational for functioning across numerous cultures.

Indeed, military members sporadically engaged in short overseas assignments following developmental training periods on cultural familiarity, face steep learning curves once immersed in the new cultural environment. High numbers of military personnel displaced on relatively short notice into geographically concentrated areas, potentially consider safety and security to be more promising when remaining among familiar military colleagues and therefore, may not engage actively with the locals. This type of response works against a dedicated commitment to enhancing any cultural awareness previously acquired. The issues of time and invested energy, versus duration and strength of PD outcomes, need to feel balanced and be perceived by learners as worthwhile. Some cultural awareness is probable; CQ is a far greater challenge. Research conducted by C. M. Vance focused on levels of CQ achievement, time in foreign countries, and engagements in cross-cultural experiences as foundational to more effective PD strategies. He identified the following effective learning-engagement activities:

- working for an international company;
- studying abroad;
- moving to a country of choice and actively seeking employment;
- international tourism;
- international internship;
- finding an international business mentor;
- teaching a mother-tongue language, e.g., English, abroad;
- government foreign service;
- multicultural teams, with members from different countries, sharing assignments;
- foreign mission/humanitarian service; and
- foreign military service.[29]

Importantly, the last option identified above is fully applicable to deployed, foreign-serving CF members, and provides significant opportunities comparable to the other strategies on the list. In the case of the CF, on-site strategies need to follow, including the integration of effective pre-site cultural awareness familiarization, training and education.

CULTURAL INTELLIGENCE PROFESSIONAL DEVELOPMENT FOR THE CANADIAN FORCES

Values-based CF leadership doctrine addresses the effectiveness of the CF as an institution in achieving mission success.[30] CQ is an inherent aspect of that effectiveness and at the same time puts predominant focus on the contexts within which the CF operates. As such, the development of CQ is informed by CF leadership doctrine, but also requires strategies for success in unfamiliar and dynamic cultural settings. The DFAIT Centre for Intercultural Learning has identified numerous cultural responsibilities and sensitivities that are critical success factors for operating in foreign circumstances with various cultures:

- recognizing the pervasive influence of cultures in all processes and activities;
- balancing leadership practices of diverse, foreign, and local people and organizations;
- achieving shared understanding of goals, objectives, roles and responsibilities;
- selecting the right people responsible for the right duties at the right times;
- maintaining consistent commitment to projects through to self-sustainability;
- setting realistic objectives with a realistic pace in schedules of implementation;
- being proactive in monitoring and managing the multicultural environment; and
- creating a high level of mutual trust among participating organizations and people.[31]

In addition, DFAIT has identified nine crucial components for the interculturally effective person, all very applicable to CF members in deployed or diplomatic circumstances:

- adaptation skills: to cope in other cultures, adjust to culture shock, enjoy host culture;
- modesty and respect: represent own culture modestly, respect other cultures, balance;
- understand culture as a concept: how culture influences people, know values clearly;
- knowledge: acquire it, possess it, expand it through history, geography, social mores;
- relationships: build strong, personal, professional, harmonious cross-cultural skills;
- self-knowledge: know own background, motivations, strengths, biases and weaknesses;
- communication: effective conveying of thoughts, opinions, ideas across cultures;

- organizational skills: improve staff morale, promote work atmosphere, be adaptive; and
- commitment: high personal and professional contributions to community, beyond work.[32]

The challenge, then, would be the selection and design of learning strategies and methodologies. As advocated in this chapter, particularly by Thomas and Inkson, and Vance, culture is a social phenomenon. As such, the development of CQ will be enhanced through the integration of training, education, experience, multicultural exposure, and assignments outside of one's own worksite/organization and cultural surroundings. Foundational information on culture, e.g., awareness, might be acquired through self-study and pedagogy-oriented acquisition through such activities as pre-deployment lectures so that basic levels of cultural awareness, and even some style and sophistication in diverse cultural circumstances, could be generated. Of course, CQ requires deeper levels of engagement with more complex understanding.

SUMMARY

A shift is required from the 20th century, Cold-War, status quo, industrial era to the dynamic and threatening 21st century – a dramatic shift toward an adult-oriented CQ model with learning strategies that emphasize the possession of knowledge and wisdom, and the mastery of high-level leader capacities through effective PD. Andragogy, with its adult learning strategies and methodologies beyond the limited all-students-same-material-group-lecturer pedagogy, provides the means to use such capacities and master such information.

As articulated by others, culture is complex, multi-layered and multi-domained.[33] Cultural wherewithal permeates the PDF leader elements. The dimensions of CQ – knowledge, cognition, behaviour, motivation and mindfulness – reflect complementary and integrated dimensions of the leader elements and major competencies that comprise the PDF. Importantly, the PDF is applicable to multiple cultural contexts and challenges from the junior and relatively straightforward level of PD to the senior level at which competent leaders have internalized ethics, values and moral reasoning that collectively reflect and create the credibility and competence of the profession,[34] otherwise understood as professional ideology. In addition, well-developed CQ and leadership reflects applied cognitive complexity, flexible and strong social/behavioural leader capacities, and mature open-mindedness and responsiveness as components of effective change capacities.

Understanding and mastering culture, possibly more than any other subject matter relevant to leaders in global, international settings, demands diverse experience and multicultural exposure, substantial interactions and personal engagements.[35] Foundational information for cultural awareness can be acquired through self-study and pedagogy-oriented instruction; however, knowledge gained through pedagogy is not sufficient to address the multiple and integrated dimensions of CQ.

This chapter has drawn together three phenomena – CQ, PD and the PDF. CQ, with all of its complexity, has been examined, described and defined in this book over six chapters, so far. Its potential to be enhanced through PD is possible but, to do so, andragogical learning strategies are obligatory in that complex phenomena require complex learning strategies far from the linear learning processes of pedagogy. The complexity of CQ – its knowledge, cognition, behaviour, motivation, mindfulness – can be addressed through integration across the five interdependent PDF leader elements and, in that context, executing PD interactively, incrementally, and progressively throughout CF members' lives and careers. The following, taken from the introduction to this volume, summarizes this challenge very well:

> It is widely recognized that cultural intelligence among all military leaders is an important contributor to mission success, whether that means combat superiority or working effectively with internal subcultures, external domestic and international organizations, or other military organizations...CQ is an integral contributor to the ability to determine adversarial intent; to work effectively across joint, inter-agency, multinational and public domains, to access and exercise whole of government approaches, and to negotiate the demands of interrelated defence, diplomacy, and development objectives.

NOTES

1 "Andragogy – The theory of teaching adults. First coined in 1833 as *Andragogik* by Alexander Kapp to describe Plato's teaching methods. Today the term is best known for the work by Malcolm Knowles." Murray Simons, *Professional Military Learning: Next Generation Professional Military Education in the New Zealand Defence Force*, (Canberra, Australia: Air Power Development Centre, 2005), xi. Today, andragogy is seen as experience-based, learner-centred, facilitator-guided, and student-active. Pedagogy is defined as the function of teachers, incorporating lecturer-centred teaching of dependent students, i.e., "empty-vessel" children as a relatively passive audience or classroom of students in receipt of "expert" presentations.

2 Michael K. Jeffery, *Inside Canadian Forces Transformation: Institutional Leadership as a Catalyst for Change* (Kingston, ON: Canadian Defence Academy Press, 2009).

3 Bernd Horn and Bill Bentley, "The Road to Transformation: Ascending from the Decade of Darkness." In Robert W. Walker (ed.), *Institutional Leadership in the Canadian Forces: Contemporary Issues* (Kingston, ON: Canadian Defence Academy Press, 2007), 1-25.

4 Ibid., 4-10.

5 John A. English, *Lament for an Army: The Decline of Canadian Military Professionalism*, (Toronto, ON: Irwin Publishers, 1998).

6 Andre Durivage and Jacques Barrette, "The Role of Values and Organizational Culture in a Strategic Planning Process," in Dan Ogle (ed.), *Strategic Planning for Police*, (Ottawa, ON: Canadian Police College, 1991), 35-74.

7 Ibid., 44-45.

8 Leonard Wong, Stephen Gerras, William Kidd, Robert Pricone and Richard Swengros, *Strategic Leadership Competencies*, (Carlisle Barracks, PA: United States Army War College, Strategic Studies Institute, 2003), 7-8.

9 Kimberly-Anne Ford and Karen D. Davis, *Cultural Intelligence, Emotional Intelligence and Canadian Forces Leader Development*. (Kingston, ON: Canadian Forces Leadership Institute & Defence Research and Development Canada, Centre for Operational Research and Analysis, 2007), CFLI Technical Memorandum 2007-01, i.

10 Robert W. Walker, *The Professional Development Framework: Generating Effectiveness in Canadian Forces Leadership*, (Kingston, ON: Canadian Forces Leadership Institute, 2006), CFLI Technical Report 2006-01.

11 For greater detail on the PDF, see Walker, *The Professional Development Framework*.

12 Bill Bentley, *Professional Ideology and the Profession of Arms in Canada*, (Toronto, ON: Canadian Institute of Strategic Studies, 2005), 110.

13 Canada. *Duty with Honour: The Profession of Arms in Canada*. (Kingston, ON: Department of National Defence, 2003).

14 See CF Leader Effectiveness Framework, Chapter 2, Figure 2-2 of this volume and Canada. *Leadership in the Canadian Forces: Conceptual Foundations*, (Kingston, ON: Department of National Defence, 2005), 19.

15 Maryanne McCarthy-Senebald, "The Development of Expertise in Maritime Surface and Subsurface (MARS) Officer Training in the Canadian Navy," (Calgary, AB: University of Calgary, 2000), Unpublished Ph D dissertation.

16 Ibid., 35, 145, 159-165.

17 Canada. *Canada School of Public Service: Your Partner in Learning*, 2006, 6. Public Service brochure available at <www.mySCHOOL-monECOLE.gc.ca>.

18 Thomas Vulpe, Daniel Kealey, David Protheroe and Doug MacDonald, *The Profile of the Interculturally Effective Person*, (Ottawa, ON: Centre for Intercultural Learning, Canadian Foreign Service Institute, Department of Foreign Affairs, Second Edition, 2001), 48.

19 Jennifer J. Bennett, "Effective Professional Development Strategies for Institutional Leaders" in Robert W. Walker (ed.), *Institutional Leadership in the Canadian Forces:*

Contemporary Issues, (Kingston, ON: Canadian Defence Academy Press, 2007), 169-194. Original research was Bennett's Master of Arts in Leadership unpublished project, Royal Roads University, 2006.

20 Ibid., 170.

21 Ibid., 171. Original source: Malcolm S. Knowles, cited in R. Hiemstra, *Lifelong Learning: An Exploration of Adult and Continuing Education with a Setting of Lifelong Learning Needs*, Third Edition. (Fayetteville, NY: Hitree Press, 2002).

22 Daniel Lagacé-Roy and Janine Knackstedt, *Mentoring Handbook*, (Ottawa, ON: Department of National Defence, 2007).

23 Jennifer J. Bennett, "Effective Professional Development Strategies for Institutional Leaders", 172.

24 Ibid., 180.

25 David Giber, Louis Carter and Marshall Goldsmith, *Best Practices in Leadership Development: Handbook*, (San Francisco, CA: Linkage Inc. and Jossey-Bass/Pfeiffer, 2000).

26 Cynthia D. McCauley, Russ S. Moxley and Ellen Van Velsor (eds.). *Handbook of Leadership Development*, (San Francisco, CA: Jossey-Bass & the Center for Creative Leadership, 1998). See also Richard L. Hughes and Katherine C. Beatty, *Becoming a Strategic Leader*, (San Francisco, CA: Jossey-Bass & the Center for Creative Leadership, 2005). The *Handbook* by C. D. McCauley et al. has numerous authors addressing the experiences, processes and issues of leader development. The Hughes/Beatty book effectively addresses the amorphous nature of strategic leadership and the challenges inherent in developing strategic and institutional leader competence.

27 Robert Kegan, *The Evolving Self: Problem and Process in Human Development*, (Cambridge, Mass: Harvard University Press, 1982). See also Daniel Lagacé-Roy, "Institutional Leader Ethics" in Robert W. Walker (ed.), *Institutional Leadership in the Canadian Forces: Contemporary Issues*, (Kingston, ON: Canadian Defence Academy Press, 2007), 109-121.

28 David C. Thomas and Kerr Inkson, *Cultural Intelligence: People Skills for Global Business* (San Francisco, CA: Berrett-Koehler Publishers, 2003). See Chapter 4, "Raising Your Cultural Intelligence", 66-73.

29 C. M. Vance, "Analysis of Self-initiating Career Path Strategies: Similarities and Differences across Genders," Paper presented at Annual Meeting of the Academy of Management, Seattle, WA, 2002. Referenced in Thomas and Inkson, *Cultural Intelligence: People Skills for Global Business*, 71.

30 Canada. *Leadership in the Canadian Forces: Conceptual Foundations*, 2-34.

31 Daniel J. Kealey, David Protheroe, Doug MacDonald and Thomas Vulpe, "Readying your Organization for International Projects," (Ottawa: ON: Centre for Intercultural Learning, Canadian Foreign Service Institute, Foreign Affairs Canada, 2004).

32 Thomas Vulpe, Daniel J. Kealey, David Protheroe and Doug MacDonald, *A Profile of the Interculturally Effective Person*, 14-19.

33 Karen D. Davis and Brian McKee, "Culture in the Canadian Forces: Issues and Challenges for Institutional Leaders" in Robert W. Walker (ed.), *Institutional Leadership in the Canadian Forces: Contemporary Issues*, (Kingston, ON: Canadian Defence Academy Press, 2007), 27-59; Kimberly-Anne Ford and Karen D. Davis, *Cultural Intelligence, Emotional Intelligence and Canadian Forces Leader Development*, 2007.

34 Bill Bentley, *Professional Ideology and the Profession of Arms in Canada*.

35 Jennifer J. Bennett, "Effective Professional Development Strategies for Institutional Leaders"; Maryanne McCarthy-Senebald, "The Development of Expertise in Maritime Surface and Subsurface (MARS) Officer Training in the Canadian Navy".

RECOMMENDED READING

Bennett, Jennifer J. "Effective Professional Development Strategies for Institutional Leaders," in Robert W. Walker (ed.), *Institutional Leadership in the Canadian Forces: Contemporary Issues*. Kingston, ON: Canadian Defence Academy Press, 2007, 169-194.

Bentley, Bill. *Professional Ideology and the Profession of Arms in Canada*. Toronto, ON: Canadian Institute of Strategic Studies, 2005.

Davis, Karen D. and Brian McKee. "Culture in the Canadian Forces: Issues and Challenges for Institutional Leaders," in Robert W. Walker (ed.), *Institutional Leadership in the Canadian Forces: Contemporary Issues*. Kingston, ON: Canadian Defence Academy Press, 2007, 27-59.

Ford, Kimberly-Anne and Karen D. Davis. *Cultural Intelligence, Emotional Intelligence and Canadian Forces Leader Development*. Kingston, ON: Canadian Forces Leadership Institute and Defence Research and Development Canada, Centre for Operational Research and Analysis, 2007. CFLI Technical Memorandum 2007-01.

Gosselin, J. P. Y. Daniel. "The Storm over Unification of the Armed Forces: A Crisis of Canadian Civil-Military Relations," In Howard G. Coombs (ed.), *The Insubordinate and the Noncompliant: Case Studies of Canadian Mutiny and Disobedience, 1920 to Present*. Toronto, ON: The Dundurn Group and Kingston, ON: Canadian Defence Academy Press, 2007, 309-343.

Jeffery, Michael K., *Inside Canadian Forces Transformation: Institutional Leadership as a Catalyst for Change*, Kingston, ON: Canadian Defence Academy Press, Forthcoming.

Thomas, David C. and Kerr Inkson. *Cultural Intelligence: People Skills for Global Business*. San Francisco, CA: Berrett-Koehler Publishers, 2003.

Walker, Robert W., *The Professional Development Framework: Generating Effectiveness in Canadian Forces Leadership*. Kingston, ON: Canadian Forces Leadership Institute, 2006. CFLI Technical Report 2006-01.

Wong, Leonard, Stephen Gerras, William Kidd, Robert Pricone and Richard Swengros. *Strategic Leadership Competencies*. Carlisle Barracks, PA: United States Army War College, Strategic Studies Institute, 2003.

CHAPTER 7

APPLYING CULTURAL INTELLIGENCE IN THE CANADIAN FORCES[1]

Major Brent Beardsley and Karen D. Davis

"It's all cultural in the end..."
Lieutenant-Colonel Ian Hope, Battle Group Commander, Task Force Orion, Kandahar (Feb–Aug 06)[2]

In the post-Cold War era, the CF has conducted, and will continue to conduct missions across the full spectrum of operations, anywhere in the world, at the tactical, operational and strategic levels, as directed by the Government of Canada.[3] These operations represent a significant shift in the current and emerging security environment and in how operations are conducted in what General (Retired) Rupert Smith has termed, "war amongst the people."[4] Support of Canadian policy direction, from preparing for operations to mission success, will continue to be the primary task of the CF. CF professionals must be firmly focused on continually improving how they prepare, how they conduct and ultimately how they will win as an armed force in future operations. This chapter is intended to provide a framework for the employment of CQ by CF leaders at the tactical, operational, strategic and policy levels, as well as across national, international, host-nation, and enemy domains.[5]

CF experience in 4th Generation Warfare (4GW) since the end of the Cold War has resulted in a continually articulated, essential operational requirement for the CF to increase the CQ capacity of its personnel and throughout the institution.[6] This requirement has been articulated as a result of various probes into the conduct of CF soldiers in Somalia in 1993 to the recent lessons learned reports on current operations in Afghanistan. For example, a 2007 review of emerging leadership issues in the CF confirmed that "cross-cultural competencies are urgently needed in conflict situations."[7] In addition, this requirement was emphasized in 2008 by virtually every senior CF leader, as well as several senior Canadian and international government representatives, who were interviewed as part of a CDS high priority investigation into coalition operations in Afghanistan.[8]

As discussed earlier, CQ is a multi-dimensional, complex, individual and ultimately organizational capacity that spans the expertise, social, cognitive, change, and professional capacities that comprise the CF PDF.[9] CQ is an essential operational requirement in the CF, and as such will be most powerful if embedded within the professional development of each and every member of the CF – from the CDS down through the entire chain of command as a command responsibility, supported by the staff and, where required, by outside experts. CQ is an essential operational requirement and the requisite CQ at the policy, strategic, operational and/or tactical level, will ensure that the primary outcome for all CF activities, namely mission success, will be achieved. This chapter examines how leaders can employ CQ to enhance the capacity of the CF in achieving this success.[10]

The purpose of the CF is the lawful and ordered application of military force in support of government policy. The general system of war and conflict, as presented in Chapter 3, provides a valuable construct for understanding how the CF accomplishes its purpose in accordance with civilian direction. Given the historical and continuing Canadian preference and practice of a limited war or conflict/conflict resolution policy employing a bipolar strategy, operational art and tactical means, only that portion of the construct will be examined in this chapter. In other words, the total or unlimited war side of the construct will not be examined.[11]

Conflict and conflict resolution have presented unique and often unprecedented challenges for military forces and other organizations seeking means of intervention to enhance socio-economic stability and security for communities worldwide. As noted by the CF Directorate of Land Concepts and Design (DLCD),

> …conflict results from a complex interaction of beliefs, actions, and structures in periods of political, economic and social instability…Accordingly, conflict engagement will require intuition to understand the essence of complex problems, ingenuity to devise innovative solutions and strength of purpose to act effectively.[12]

Insofar as CF leaders have a responsibility to coach, mentor, develop, and guide, they are key influencers of the development of CQ among individuals and their teams as a whole. A CQ-effective team will be better prepared to meet the challenges of military operations within the context of national, international, host-nation and enemy domains.

Importantly, CF members who recognize the importance of CQ will seek opportunities to develop their CQ in a quest to understand the cultural lens through which various government and non-governmental elements of international communities, Canadians and the enemy view and interpret the world and more importantly, how they will view and likely respond to CF initiatives and behaviours, both at home and around the world. Further, CQ starts with the conscious and consistent understanding of the centrality of culture to the mission capability of the CF.

EMPLOYING CULTURAL INTELLIGENCE AT THE POLICY LEVEL

Policy development and endorsement in Canada is dominated by the Cabinet, on behalf of the Government of Canada, including the sole and ultimate responsibility for determining policy objectives for the CF.[13] CF Leaders at this level act as advisors, usually in close collaboration with senior advisors from the Department of National Defence (DND), OGDs or non-governmental advisors, to the policy decision-makers in Cabinet.[14] "Good policy" development is dependent upon "good advice" informed by timely, relevant and accurate information.

Advice to policy-makers is usually developed by the Assistant Deputy Minister – Policy (ADM (Pol)) in collaboration with the CDS Strategic Joint Staff (SJS). The CDS is ultimately responsible for ensuring that policy advisors and decision-makers access appropriate input. The CDS and the executive institutional leadership of the CF cannot become experts in every culture with which the CF interacts; however, they will rely upon a high level of institutional CQ. That is, the effectiveness of the CF relies upon the ability of senior leaders to understand and negotiate complex systems, successfully develop and implement strategy, create an adaptable and agile learning organization, and successfully facilitate change within the context of national strategic culture and international security and defence challenges.[15]

Citing the increasing complexity of the policy-making environment, strategic planning expert Carl Taylor claims that the traditional policy cycle, presented in Figure 7-1 below, is not sufficiently agile and adaptable.

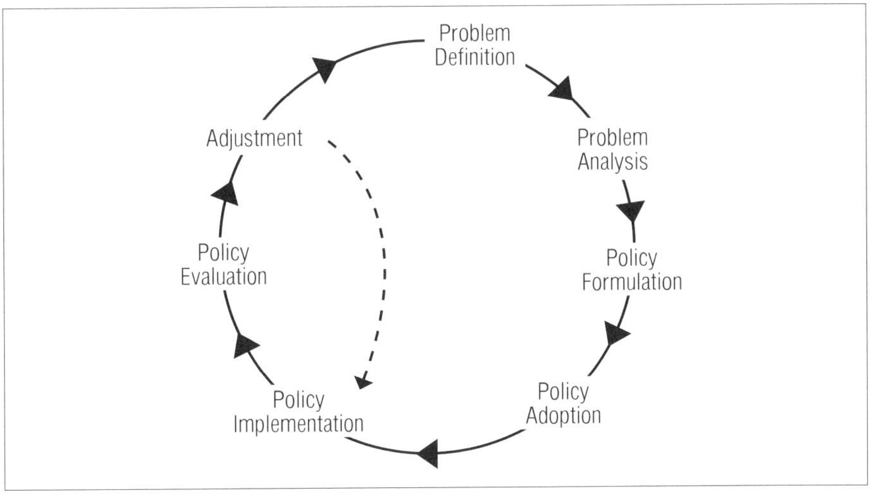

FIGURE 7-1: TRADITIONAL POLICY CYCLE[16]

The ACIDD Test: An Alternative Policy Planning Framework

Alternatively, Taylor proposes the ACIDD (Analysis, Choice, Implementation, Debate and Decision) Test as a policy planning framework. Reliance on a policy framework such as this is not only suitable for policy development in an increasingly complex world, but demands CQ as an integrated part of the policy process. Indeed, considerations integral to CQ have informed the development of this framework. As he notes:

> Diversity in values, beliefs, cultures and behaviours makes reaching a consensus more difficult, while the internationalization of problems and the growth of multinational agreements and instruments have tended to reduce the capacity for independent action by individual nations. Both the instruments and processes for solving problems have changed. It is against this background that policies will be developed and adopted.[17]

Analysis

The ACIDD model, presented in Figure 7-2 below, considers, from the initial policy development or analysis stage, the relationship between issues and their impacts, provides policy-makers with information such as cause(s), duration, nature and scope of the issue or challenge, and information about the affected populations and their characteristics.

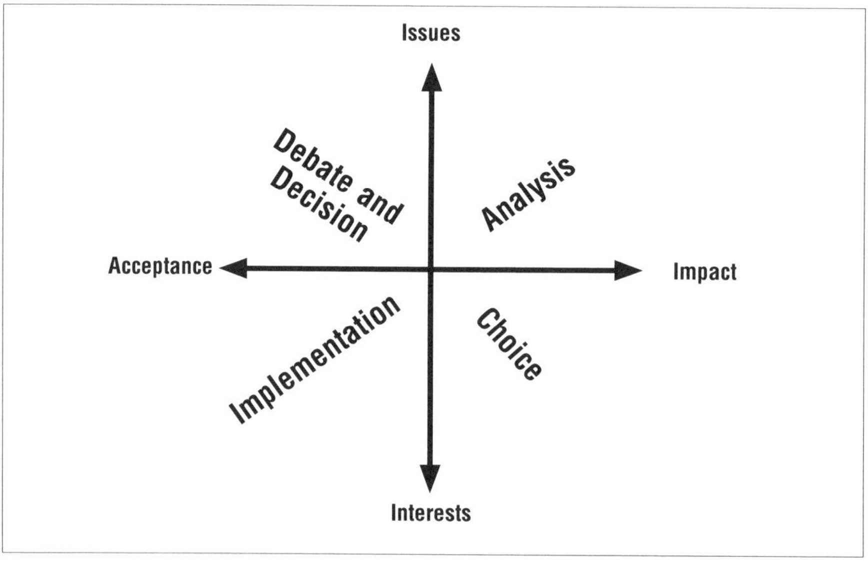

FIGURE 7-2: THE ACIDD TEST: AN ALTERNATIVE POLICY PLANNING FRAMEWORK[18]

Choice

The choice stage defines the problem based on qualitative and quantitative information from all four quadrants of the model, thus reflecting a thorough understanding of the issues, their impacts, and their determined and potential acceptance by national and international interest groups, as well as impact on host nations and the enemy or belligerents.

Implementation

Implementation is essentially the integrated "how" of the policy development process, dealing with operational issues such as affordability, feasibility, tools and delivery mechanisms. This phase is distinct from the traditional policy cycle in that it also includes "design" considerations to facilitate acceptance by interest groups and stakeholders such as OGDs, the Canadian public, and the host nation.

Debate and Decision

The debate and decision process explicitly considers politics and political risk. In addition to traditional considerations of effectiveness and efficiency, the ACIDD model includes considerations related to the political effort that will be required to adopt, promote and implement the policy. As emphasized by Taylor, application of this model requires "astute political leadership and vision;"[19] that is, well-developed CQ among senior CF leadership.

Policy Development in the Canadian Forces

The CDS sets the standard by identifying requirements and determining the priority of resources that will be applied to a given challenge. While all staff officers will have some degree of CQ, specific cultural expertise may be outside of their knowledge or experience. The Intelligence Branch typically takes the lead in collecting and analysing culture-specific information; however, this is not the sole source of CQ. As the CF continues to enhance its CQ capacity and capabilities, it must inevitably permeate the entire institution and is particularly important for those members selected for employment in the SJS.[20]

In their capacity as advisors at the policy level, CF leaders must base their advice, recommendations, options, etc. on timely, relevant and accurate information. This information must be informed by CQ and integrated into the policy development process using a non-linear process such as soft systems methodology (introduced in Chapter 1), or a policy development tool such as the ACIDD Test. Sound advice depends upon determining the information senior leaders will require to make a sound assessment; that is, collecting cultural information, subjecting it to analysis within each of the four domains and then disseminating the pertinent information as directed and required.[21]

First and foremost, CF policy advisors must understand the strategic culture of our nation (as discussed in Chapter 3) before they can even attempt to understand the strategic culture of others[22] (e.g. our international partners and sponsoring organizations, non-governmental players and our adversaries).

In an ideal world, the potential CF deployment areas are identified well in advance, and extensive cultural studies and preparatory professional development are conducted (e.g. language training, etc) to ensure the requisite CQ is widely available. The CF cannot realistically possess the requisite CQ for every area of the world. In addition, given the uncertainty of the current and emerging security environment, the CF can be deployed with little advance warning to an area about which the requisite CQ is not readily available within the institution. CF leaders acting as advisors at the policy level must acquire cultural advisors with a high level of knowledge and experience pertaining to the potential deployment area. It is important to recognize that cultural advisors are not a substitute for CQ among CF leaders. However, cultural advisors will frequently be essential to the development of culture-specific awareness among receptive, open-minded leaders who are enhancing their CQ acumen for the mission at hand.

Cultural Advisors

Cultural advisors at this level may be recruited from several sources. Ideally, institutional leaders will have an extensive network of contacts throughout government, the private sector, academia and various diasporas and communities. Potential CQ experts may be currently serving in the CF, both Regular and Reserve components, or the DND, and can be tasked to support the SJS throughout the policy development process. Personnel do not have to have a direct ancestral link to the culture of the potential deployment area to be considered experts; they may have lived, served, studied or travelled extensively in the area. If this expertise is not resident in the CF, Canadians serving in OGDs such as DFAIT, CIDA, etc., who have extensive experience in the potential deployment area can also be a valuable resource. Canadian academia, especially scholars specializing in the area, and Canadian residents who are members of, or have connections within a diaspora are also potential advisors. Ideally, a mix of cultural advisors – CF members and civilians, Canadians and other members of diasporas – should be consulted to generate a range of advice and expertise for the CQ-sensitive SJS and to begin the process of enhancing CQ within those CF staff members who will be working the operation. The effective development and use of networks is essential to recruiting the best advisors available.

Identifying Information Requirements

Once advisors have been engaged, the next step is to identify the information required to build the necessary advice to government. For example, information in reference to the geography; history; role of ideology and religion; culture; governance, etc. will be necessary to provide an understanding of the culture of the groups within the potential deployment area. The input and advice of cultural advisors will be critical to ensuring that the right questions are asked and that sufficient accurate information sources are available in reference to all four domains. Understanding how potential Canadian policy will be received in the four domains and what words and actions could be taken to effectively influence that reception are critical at this level.[23]

Once the information requirements have been identified, cultural advisors and staff officers possessing high levels of CQ will create an information collection plan. Sources like literature reviews, in-theatre information collection, consultations with academic experts and research centres, the internet, reviews of international, host-nation and enemy media, interviews with subject matter experts and especially the advice and contacts of cultural advisors must be used to collect the necessary information.[24]

The macro information must then be submitted to analysis. The analysis of the cultural information must be a multi-disciplinary activity with cultural advisors providing the expertise, especially in terms of evaluating motivation and expected behaviours. Once the analysis is complete, it must be integrated where appropriate in the policy development process. The senior CF leaders who present the advice to government must be confident that the advice, options, recommendations etc. have been built on a solid foundation of cultural expertise and that the impact of culture has been thoroughly assessed and assisted in the development of CF input to policy-makers.[25]

Policy Direction and Communications

Policy direction will always have a communications annex which will provide the key ideas, talking points, messages etc., that the government believes are critical to national, international and host-nation support and of the greatest challenge to the enemy. This communications annex forms the basis of the common narrative that will be further enhanced at each subordinate level. In other words, all levels must be "singing to the same sheet of music", not only in the CF, but within OGDs and ideally within NGOs, international organizations and allies, and the host nation and its agencies.[26] Although beyond the scope of this paper, information operations, as a potentially war-winning or war-losing capability, are also enhanced by CQ.[27] The success of these operations will largely be based on how effectively CQ has been integrated into the communications directive.

The final step in the preparation of advice at the policy level is to disseminate the information throughout the SJS and to subordinate levels in the form of briefings, presentations, literature, educational and training materials, recommended self-development study packages, etc., which will enhance CQ in relation to specific missions and tasks. This stage of the process will also begin to form the basis of information collection for the operational planning process and for the development of requirements for training, personnel selection, in-theatre information collection, contracting of cultural advisors/interpreters, etc.

Once the government has made the policy decision, the CF will focus on the development and management of the military strategy as part of the national strategy to achieve the policy objectives of government. From this macro level of policy direction, the development of CQ has only begun and must become a more intense, multi-dimensional, multi-player activity with a specific focus on the behavioural and motivational elements within the four domains.

EMPLOYING CULTURAL INTELLIGENCE AT THE STRATEGIC LEVEL

The development of a military strategy is a component of greater national and most likely international mission strategy. At the strategic level, leaders seek to match military ways and means with other national ways and means, begin the shaping of the external environment, and improve and sustain the internal environment to achieve the ends or objectives of policy; the development and management of the strategic directive is the ultimate outcome. Armed with policy direction, the focus of the SJS will be defined as it determines, with a "whole of government" approach and in conjunction with our allies, the ways and means to achieve the policy objectives.[28] The use of a bipolar strategy, as discussed in Chapter 3, will necessitate the subordination of the military or battle pole to the non-military or non-battle pole.

The development of the strategy must be informed by CQ. Cultural information requirements will continue to be identified, collected, analysed and disseminated up and down the chain of command. CQ in the SJS will be further enhanced through the advice, expertise and assistance of cultural advisors. Ideally the same advisors and same CF staff employed in developing the advice to policy-makers will continue to be employed in the development and management of the strategic directive. In addition, non-linear approaches such as those discussed in Chapter 1 and inherent within models such as the ACIDD test will inform the most effective strategy.

Cultural Overlays

At the strategic level, the research into the four cultural domains must be broadened. The development of cultural overlays, which identify cultural groups, sub-groups and networks; the social and power structures; gender and diversity factors, and leaders and the means they use to influence culture, especially the behavioural and motivational elements, can assist in the process. For example, the SJS must try to determine how cultural knowledge and social psychological processes influence different demographic and social components of the population; how to build strategic relationships with these various components; how to communicate across cultural lines; how to acquire the necessary language abilities; and how to manage the perceptions of various groups toward members of the CF and other external groups. Importantly, appropriate mechanisms must be put in place to continuously develop CQ, as well as disseminate what is learned to other leaders who will be involved in the mission.

Once the strategic directive has been developed, approved and issued, it will be managed by the SJS who will continue to support subordinate levels with CQ input as required during the five stages of a mission (tactical level): warning, mounting, deployment, employment and redeployment, all discussed in greater detail later in this chapter.

EMPLOYING CULTURAL INTELLIGENCE AT THE OPERATIONAL LEVEL

Responsibility for the development of the campaign plan at the operational level rests with Canadian Expeditionary Force Command (CEFCOM) for missions outside of Canada, and with Canada Command (CANCOM) for missions conducted within Canada. The development and management of the campaign plan for the mission will be a collaborative whole of government effort and most often requires integration into a wider international or coalition plan.[29] Importantly, campaign planning and a systems approach are not mutually exclusive; that is, the fusion of CQ with a systems approach, as suggested in Chapter 1, will result in a more holistic, and thus effective, campaign plan.[30]

Campaign Planning and Enhancing the Cultural Overlay

The planning process currently employed by the CF is the operational planning process (OPP), a critical early stage of which is the intelligence preparation of the battlespace (IPB).[31] The IPB provides optimum opportunity to include culture in campaign design, thus combining the increased complexity of asymmetric and coalition warfare. The OPP is only one approach to operational planning. Increasingly the CF is recognizing the advantages of systems-based approaches to planning and have begun to employ systemic operational design (SOD).[32] SOD, which has a cultural component, makes it possible to learn and adapt from cultural tensions.[33] Regardless of the design or planning approach, the fusion of culture and systems analysis is an important inclusion at this critical juncture between strategic objectives and tactical level operations.

While the area of operations (AO) may be clearly defined by boundaries on a map, the free flow of people and information across boundaries and borders, while always a factor, is an increasingly dynamic consideration in a globalized world. Increased specificity in the cultural overlay seeks clan, tribal, ethnic, religious and other links from within the AO to the area of interest (AI). In addition, information related to communication and media links, external economic and political ties, as well as distributed financial, moral and logistics links to cultural groups within the AO is critical. Comprehensive intelligence embedded within the host nation is paramount; however, it is equally important to map host-nation processes to the international and enemy domains as part of the IPB process.[34]

Cultural advisors and operational planners with optimum levels of CQ must also acquire information on the following, each discussed in greater detail below, to enhance their operational planning:

- the society;
- the social structure;
- the cultural specificity;
- the language;
- the power and authority; and
- the interests.[35]

Society

The concept of society is difficult to define. As presented in Chapter 1, society can be defined as a population whose members are subject to the same political authority, occupy a common territory, have a common culture and share a sense of identity. In reality, a society does not require all of the above factors and one or more may be sufficient to constitute a society within the AO. No society is homogeneous. There may be a dominant culture or there may be a number of cultures living in some form of coexistence. At best, describing the society in the AO is a starting point for enhancing CQ.[36]

Social Structure

Within the social structure, the groups (whether racial, tribal, ethnic, religious or another category) must be identified in terms of any external links with the AI; the national and international domains must also be determined and evaluated. Once the groups have been identified, the relationships between groups, whether formal (e.g. long-standing alliances), or informal (cohabitation or friction), and divisions (long historical conflict) between groups and cleavages (e.g. religion) that cut across group lines must be determined. In addition, every group within a social structure has a network whether it is between elites, areas, gender sub-groups, wider ethnic or religious links within the host nation and possibly beyond, reaching into the national and international domains. Careful analysis of this information provides more depth to the cultural overlay.[37]

Institutions are committed to a common task (e.g. education institutions) and groups engage in patterned activity within those institutions to achieve it. Institutions are the basic building blocks of a society and cut across group lines. Identifying key institutions within an AO, along with their connections to other institutions in the national and international domain, provides critical information on centres of gravity for inter-group co-operation and social development.[38]

While organizations are a form of institution, they restrict membership, define goals, conduct an activity, have fixed locations and have financial and logistical support. Communications or media organizations, religious organizations and places of worship, economic organizations such as factories, and social organizations like sports teams, are just a few examples of the types of activities around which organizations are developed. At the operational level, all types of major organizations need to be mapped so that detailed locations can be determined at the tactical level.[39]

The final aspects of social structure that must be examined are the roles and statuses within the social structure and the social norms that guide conduct. Social and economic status brings with it a specific role which guides how a person speaks, thinks and acts. Interaction between CF personnel and individuals in various status positions must be informed by the nature of the individuals' status and the role they are expected to play. Social norms include the expected reaction to the killing of a family member or relative, appropriate treatment of women, gift giving, etc. All possible social norms must be identified to ensure CF personnel are aware of, and ideally understand, how they will impact the mission within the AO.[40] Pashtun tribal mores (Pushtunwali), for example, demand that all male relatives take up arms to defend fallen brothers, sons, and fathers. The Soviets learned this lesson in Afghanistan in the 1980s after they killed over a million Pashtuns, only to increase the number of guerrillas that they faced at the end of the war.[41]

Once the social structure has been mapped, operational-level planners and their cultural advisors need to conduct a deeper and more specific analysis of culture than was done at the macro-policy and strategic levels. If the social structure is the skeleton of a society then culture is the integration of muscle and flesh that envelops and provides critical support to the structure. Individuals and teams that have developed effective CQ will have the capacity to gain a fuller understanding of how the various elements of the culture, both the visible and the less tangible, reinforce and support one another or alternatively, have a relationship based upon real or smouldering conflict.

Cultural Specificity

In analysing the culture of the host nation, enemy or international domains, identities and identity relationships are important. As discussed in Chapter 5, identity and identity development provide clues to how one makes meaning of the world around him/her, and thus how individuals and communities choose to engage with other individuals and communities. Every individual has several identities but typically one is dominant in terms of influences on values, beliefs, motivations, and behaviours. For example, religious identity may be a more significant influence than national, tribal or ethnic identity, or clan identity may be paramount. Determining the primary and secondary identities within the groups in a cultural overlay is critical to understanding how individuals and groups may respond in various situations, including the extent to which they will support or challenge CF and coalition forces.[42]

Within the identities determined in the analysis, the beliefs, concepts and ideas held to be true must be identified. These beliefs can be core (firmly held as absolute like a belief in God), intermediate (usually defined by opinion leaders and somewhat susceptible to change over time), or peripheral beliefs that are open to debate. Determining core beliefs in the four domains is essential in ensuring that information operations do not unnecessarily confront or threaten core beliefs; in determining how and the extent to which intermediate beliefs can be influenced and by whom; and identifying the peripheral beliefs which can be debated logically and changed, if required, to support the mission.[43]

Of even more importance than beliefs are values that are relatively enduring and thus have a significant influence on behaviour. Again, values are largely derived from and consistent within somewhat definable groups. In particular, those in conflict with Canadian values or in conflict with those values predominant within the host nation need to be determined. In addition, the values of the enemy, our own nation and any international partners need to be identified and considered in the planning process.[44] In particular, it is important to recognize those values which unconsciously inform the everyday beliefs and behaviours of CF and coalition members and that may have an unanticipated impact on relationships with other organizations, host-nation communities, international partners, and belligerents.

In addition to beliefs and values, attitudes and perceptions must be identified for all of the groups in the four domains. Affinities or aversions between groups can provide key information in relation to host-nation networks, relationships, attitudes and perceptions that are susceptible to change and thus potential targets for information operations.[45]

Language

One of the most important information requirements at the operational and tactical levels is language. Identifying the languages in use in the AO and AI, and determining which language to use with the national, enemy and international domains are critical requirements. The technical use of language can be learned; however, determining appropriate social use of language requires a high degree of CQ. Oral expression and written text are the most obvious forms of language; however, language is integrated within a broader system of communication including, for example, symbols, "body language," and patterns (linear and non-linear) of activities and decision-making. Understanding the roles and use of language will ensure those personnel with valuable language skills or the ability to acquire language skills, are effectively employed in the operation.

Power and Authority

Once these culturally-specific skills, knowledge and attitudes have been obtained and assessed, mapping of the cultural overlay can continue by determining the nature of power within the host-nation society, and the enemy, international and, where appropriate, national domains. Determining formal or informal power holders, especially in the areas of central and local government, political parties and interest groups, government agencies etc. permits the mapping of power. Power can be coercive, social, economic, rational-legal based, charismatic or traditional authority. Knowing who holds what type of power and how they employ it provides the insight into how to support, reinforce, change or deny that power as required in the campaign plan to achieve mission success.[46]

Interests

Interests are the core motivations that drive behaviour. Determining the interests of groups in the host-nation, enemy, national and international domains will ensure that CF operations are conducted to either neutralize the interests of certain groups such as the enemy or components of the enemy (e.g. popular public support), or to enhance the interest of other groups such as the host-nation government or a social movement such as a peace process or democratization movement.[47] This can only be accurately done if all of the previous information requirements have been met and effectively analysed.

As the campaign plan is being developed with a "whole of government" and coalition approach, it is important that CF leaders continue to apply a high level of CQ in ensuring that the campaign plan is culturally informed, adaptable, and effective in theatre. The use of cultural advisors and CQ-skilled operational planners, open to information from all sources, will ensure that a culturally competent campaign plan is developed and maintained.

EMPLOYING CULTURAL INTELLIGENCE AT THE TACTICAL LEVEL

The tactical level is where "the rubber hits the road." This is where battles are won or lost and where national and international policy, military, national and coalition strategy, and the campaign plan succeed or fail. This is where the words, gestures and actions of all CF personnel engaged in the mission area will directly impact mission success or failure. As illustrated by the case study at Appendix C, this was evident in the case of CF operations at Oka in 1990. CF leaders at the tactical level must be prepared to adapt their behaviour to culturally-based activity to effectively influence real and perceived mission success.[48]

The CF, as an institution, is well positioned to enhance its CQ on operations at the tactical level. As discussed in Chapter 4, an increasingly diverse CF provides growing potential for the development of CQ. The extent to which diversity is effectively facilitated before and during deployments will impact the strength of the team's CQ in theatre. In addition, as CF personnel gain increased operational experience living, moving, communicating, operating and fighting in culturally diverse environments, their level of CQ will only be enhanced by each of these experiences. As the CF continues to deploy to ever varied locations and as staffs continue to work through contingency operations, CQ will only grow. Outside the CF, the institution can take advantage of an ethnically and culturally diverse population, a world-class education and research system, and the wealth of existing knowledge, skills and attitudes within our society to further enhance its CQ.

Missions at the tactical level are conducted in five stages, each of which is discussed briefly below: warning; mounting; deployment; operations; and rotation/redeployment.

Warning Stage

Upon receiving the initial warning of an upcoming mission, mechanisms and processes should be immediately activated to accelerate the further development of CQ across the knowledge and behaviour dimensions of the tactical component. Those members of the CF identified for deployment should gain an early appreciation of the relevance of CQ to their upcoming mission and be given access to information and resources to enhance their knowledge and capacity in reference to the cultural aspects of the anticipated mission. Initially, this information may be very general and macro in nature but the push of information can still begin the process of applying CQ to the mission at hand.

Mounting Stage

During the mounting stage, the development of CQ within the tactical component will be broadened and deepened as more information becomes available, training programs are developed and implemented, and training resources such as literature, presentations and cultural advisors are provided to the tactical component. While leaders at the tactical level will require high levels of CQ, all personnel need as much cultural information as can be provided. As a minimum, CF personnel must understand the importance of CQ in the achievement of their mission. This will enhance motivation to further develop knowledge and cultural expertise. Historically, the CF has focused on mission-specific, behavioural "dos and don'ts" to avoid offending individuals within the host nation in particular. Ideally, CQ development among members of CF tactical units will be comprehensive and inclusive of all domains of CQ: motivational, cognitive, knowledge, and behavioural development. Pre-deployment training and preparations that enhance CQ are important to all CF members, and in particular to those who will assume leadership positions in tactical roles. Cultural advisors, instructors and role players, for example, provide added value to pre-deployment training activities and exercises, as they provide the opportunity for each member to interact, receive feedback and develop knowledge to enhance their culture-specific awareness as well as overall CQ development.

Deployment and Operations Stages

During the deployment and operations stages, tactical commanders at the highest levels require Canadian and host-nation cultural advisors to provide timely, relevant and accurate information and advice to the commander and his staff as they plan and execute their operations. Their presence and availability will ensure that the CQ of the commanders and their staff is in a constant development and enhancement mode. Although cultural advisors may not be readily available, most tactical elements will have a locally engaged interpreter who can be employed as a cultural advisor and contribute to the continuous development of CQ among members of the tactical unit. As experience using CQ grows, so too will the CQ of all within the tactical component. The effective employment of CQ at the tactical level will ensure that every word, gesture or action made by a member of the CF contingent contributes to mission success and does not imperil or detract from that objective. Some cultural information provided may prove to be false or no longer relevant. A dynamic in theatre lessons-learned system can ensure that this learning is returned to the nation for future reference and training.[49]

Redeployment Stage

During rotation or redeployment, CQ lessons learned must be provided to the incoming rotation or up the chain of command to further enhance our cultural expertise, not only as individuals but as an institution. In addition, CF members coming out of theatre who possess an optimum level of current experience and knowledge should be encouraged to share that information in formal and informal venues. Such knowledge-sharing processes are essential activities across all levels, and not the privileged domain of a few. Ideally, contributors and participants in information and discussion sessions will be solicited and made available from multiple perspectives and roles (rank, trade, role in theatre, etc.), thus ensuring a broadly inclusive process of learning across the CF.

SUMMARY

This chapter has provided a brief overview of how CF leaders can effectively employ CQ on operations. Relying upon the general system of war and conflict at the policy, strategic, operational and tactical levels, CQ has been applied from advice for the formulation of policy, to the development of national and military strategy, to the design of campaign plans, to planning and conducting operations at the tactical level. CQ must permeate all planning and execution tasks and be applied to the national, international, non-governmental and enemy domains. In addition, CQ and more importantly the effective application of CQ, is a command, not a staff, support or advisory competency and responsibility. While experts from a wide range of sources may be utilized to provide mission area-specific advice, information and explanation, the responsibility of ensuring that policy, strategy, operational art and tactics are fully informed by CQ rests with leadership from the top to the bottom of the CF.

The greatest hurdle that the CF will have to overcome as it seeks to professionally develop CQ in all of its members is changing the misconception that CQ is just another "nice to have skill," "an afterthought" or an item on a checklist that a series of simple, basic lectures or a pocket card can address. It will require leaders at all levels to recognize and internalize that effective CQ is an essential operational requirement that will directly contribute to mission success and may in many operations prove to be the factor that will bring victory or defeat.

NOTES

1 The basis of thought informing this chapter can be found in John A. Lynn's seminal work on war and culture, *Battle: A History of Combat and Culture*, (Cambridge: Westview Press, 2003).

2 Lieutenant-Colonel Ian Hope was the first Canadian to become the Battle Group Commander of Task Force Orion, Kandahar, Afghanistan. He gave a presentation to the staff of the Canadian Forces Leadership Institute upon his return from Afghanistan in 2006 and made this statement while providing numerous examples of CQ based on his experience in Afghanistan, thus inspiring the ongoing work for this volume.

3 Canada, *Duty with Honour: The Profession of Arms in Canada*. (Kingston: ON, Canadian Defence Academy, Canadian Forces Leadership Institute, 2003), 12-23.

4 Rupert Smith, *The Utility of Force: The Art of War in the Modern World*, (London: Allen Lane, 2005). Smith's thesis is that war has evolved in the current and future security environment to what he terms" war amongst the people."

5 For discussion of these domains, see Emily Spencer, *Crucible of Success: Applying the Four CQ Domain Paradigm, The Canadian Forces in Afghanistan as a Case Study*, (Kingston, ON: Canadian Forces Leadership Institute, Technical Report 2007-05).

6 Thomas Hammes, *The Sling and the Stone: On War in the 21st Century*, (Osceola, WI: Zenith Press, 2004). Hammes provides an excellent examination of the evolution of the four generations of warfare and is credited with one of the best descriptions of 4th Generation Warfare and the challenges it will provide to Modern Western militaries.

7 Commander R. S. Edwards, *Leadership in Canadian Forces Operations: Emerging Issues and Future Trends*, (Kingston, ON: Canadian Forces Leadership Institute Technical Report 2007-07, 2007).

8 Canada, *Broadsword or Rapier? The Canadian Forces Involvement in 21st Century Coalition Operations*. (Kingston, ON: Canadian Forces Leadership Institute, 2008). The research and analysis in this report identified CQ as a critical issue/common theme and provided a number of recommendations on how the CF could improve CQ to improve operational effectiveness on coalition operations.

9 Robert W. Walker, *The Professional Development Framework: Generating Effectiveness in Canadian Forces Leadership*, (Kingston, ON: Canadian Forces Leadership Institute Technical Report 2006-01, 2006).

10 For a detailed description of the CF Effectiveness Framework, of which mission success is the primary outcome, see Canada, *Leadership in the Canadian Forces: Conceptual Foundations*.

(Kingston, ON: Department of National Defence, Canadian Defence Academy, Canadian Forces Leadership Institute, 2005); Canada, *Leadership in the Canadian Forces: Leading People*, (Kingston, ON: Canadian Defence Academy, Canadian Forces Leadership Institute, 2007); and Canada. *Leadership in the Canadian Forces: Leading the Institution*. (Kingston, ON: Canadian Defence Academy, Canadian Forces Leadership Institute, 2007).

11 For more detail on the Canadian way of war, see Dr. Bill Bentley, "Canada's Way in War," p. 83-97 *Institutional Leadership in the Canadian Forces: Contemporary Issues*, Robert W. Walker (ed.), (Kingston, ON: Canadian Defence Academy Press, 2007).

12 National Defence, *Land Operations 2021, Adaptive Dispersed Operations: The Force Employment Concept for Canada's Army of Tomorrow*, (Kingson, Ontario: Directorate of Land Concepts and Design, 2007), 6-7.

13 Canada, *Leadership in the Canadian Forces: Leading the Institution*. 41-59.

14 Glen Milne. *Making Policy: A Guide To The Federal Government's Policy Process*. (Ottawa: Queens Printer, 2004). This straightforward manual provides an excellent guide as to how policy is made in the federal Government of Canada.

15 See Canada, *Leadership in the Canadian Forces: Leading the Institution*, for further discussion of these processes in the CF.

16 Adopted from Carl Taylor, "The ACIDD Test: a framework for policy planning and decision making," *Optimum, The Journal of Public Sector Management*, 27 (4, 1997), 54.

17 Carl Taylor, "The ACIDD Test," 54.

18 Adapted from Carl Taylor, "The ACIDD Test," 55.

19 Carl Taylor, "The ACIDD Test," 57.

20 Emily Spencer, *Crucible of Success: Applying the Four CQ Domain Paradigm. The CF in Afghanistan as a Case Study*, (Kingston, ON: Canadian Forces Leadership Institute Technical Report 2007-05, 2007).

21 Canada, *Joint Intelligence Doctrine* [B-GJ-005-200/FP-000], (Ottawa: CFD, 2003). Chapter 2 of this keystone CF manual provides the doctrine for the direction phase of the intelligence cycle.

22 In Chapter 3 of *Leadership in the Canadian Forces: Leading the Institution*, the relationship between strategic culture and strategy is described and exemplified with a case study (41-44).

23 Canada, *Canadian Forces Operations* [B-GJ-005-300/FP-000], (Ottawa: Canadian Forces, 2005) is the capstone CF doctrine manual for operations. Chapter 1, Section 1 describes the National Policy Framework within which the CF operates. Glen Milne, in *Making Policy*, Chapter 4 (39-46) describes how policy proposals are developed.

24 Canada, *Joint Intelligence Doctrine*, Chapters 2 and 3 describe the information collection process.

25 Glen Milne, *Making Policy*, Chapters 4 (39-46) and 5 (47-52) describe the development of policy proposals and Cabinet decision-making.

26 Glen Milne, *Making Policy*, Chapter 5 (53-58) describes how policy is implemented.

27 See W. N. Peters, *Shifting to the Moral Plane: The Canadian Approach to Information Operations*, (Kingston, ON: Canadian Forces Leadership Institute Technical Report 2007-04, 2007), for an overview of the state of information operations in the CF and observations in reference to the linkage between cultural intelligence and leadership.

28 Bernd Horn, (ed.), *Contemporary Issues in Officership: A Canadian Perspective*, (Ottawa: Canadian Institute of Strategic Studies, 2000). In Chapter 6 (145-180) of this work, Dr. L. W. (Bill) Bentley provides an excellent essay that began his work on the model of the General System of War and Conflict. Within this Clausewitz-based model, there are two types of strategy as described in Chapter 3 of this book. The analysis in this chapter has been based solely on the preferred Canadian use of a bipolar strategy to achieve a limited policy objective. A strategy of annihilation has not been employed by Canada since World War II and even began to lose favour during the Cold War. The use of a bipolar strategy remains the preferred Canadian approach to strategy. Canada, *Canadian Forces Operations*, Chapter 1 provides CF doctrine for the Strategic Level in the CF.

29 Canada, *Canadian Forces Operations*, Chapter 2, Section 3 provides the CF overarching doctrine for campaign planning. Canada, *The Canadian Forces Operational Planning Process*, (Ottawa, ON: Canadian Forces, 2008) provides the current CF doctrine for the use of the operational planning process. In Chapter 2 the doctrine for campaigning, campaign design, planning, conduct and contents is defined.

30 See also Kent O. Strader, *Culture: The New Key Terrain Integrating Cultural Competence into JIPB*, (Fort Leavenworth, Kansas: School of Advanced Military Studies, United States Army Command and General Staff College, Monograph AY 05-06, 2006).

31 Canada, *Joint Intelligence Doctrine*, Chapters 2 and 4 provides the CF doctrine for the use of Intelligence Preparation of the Battlespace.

32 A number of alternatives to OPP are being developed and experimented with by allied forces. The Israelis and the US Marines have done work into a new approach to operational planning termed "Systemic Operational Design." This approach has its roots in modern complex systems theory. However, the CF has not endorsed any other approach than the OPP but must stay abreast of the development of other and perhaps better approaches to planning.

33 Kent O. Strader, *Culture: The New Key Terrain Integrating Cultural Competence into JIPB*, 47.

34 Canada, *Joint Intelligence Doctrine*, Chapter 4 provides the CF doctrine for the mapping of the international and enemy domains as part of the IPB process.

35 United States. *The U.S Army and Marine Corps Counterinsurgency Field Manual*. (Chicago: The University of Chicago Press, 2007). 84-100. This current US military doctrine manual, based largely on their experience in Iraq, Afghanistan and as a result of a major high level research project led by General David Petraeus (subsequent US Force Commander in Iraq), General James Amos (USMC) and expert scholar Lieutenant-Colonel John Nagl provides the most clearly articulated examination of the role of CQ in counter-insurgency operations. The examination is equally applicable to any other low to mid intensity conflict across the spectrum of operations. This section provides definitions for many terms used in any discussion of CQ.

36 Ibid., 85

37 Ibid., 85-87.

38 Ibid., 87.

39 Ibid., 87-88.

40 Ibid., 88-89.

41 Thomas Johnson, "Tribal Politics: Why we must understand the human terrain," *Vanguard*, (May/June 2008), 12-14.

42 United States, *The U.S Army and Marine Corps Counterinsurgency Field Manual*.

43 Ibid., 91.

44 Ibid., 91.

45 Ibid, 91-92.

46 Ibid., 94-97.

47 Ibid., 97-100.

48 Canada, *Canadian Forces Operations*, Chapter 4 provides the overarching doctrine for the employment of CF on operations at the tactical level. Chapter 5 examines the use of force. Chapter 6 examines collective training. Chapter 7 examines Task Force organization and Chapters 8-13 examine specific types of operations.

49 At the time of writing the CF is creating a comprehensive CF Lessons Learned Organization and Process.

RECOMMENDED READING

Canada. *Canadian Forces Operations [B-GJ-005-300/FP-000]*. Ottawa, ON: Canadian Forces, 2005.

Canada. *Joint Intelligence Doctrine [B-GL-005-200/FP-000]*. Ottawa, ON: Canadian Forces, 2003.

Canada. *The Canadian Forces Operational Planning Process [B-GJ-005-500/FP-000]*. Ottawa, ON: Canadian Forces, 2008.

Canadian Forces Leadership Institute Project Team, *Broadsword or Rapier? The Canadian Forces' Involvement in 21st Century Coalition Operations: CDS Critical Topic Number 6 Project Report*. Kingston, ON: Canadian Forces Leadership Institute, Technical Report 2008-01.

Hammes, T. *The Sling and the Stone: On War in the 21st Century*. Osceola, WI: Zenith Press, 2004.

Lynn, J. *A History of Combat and Culture*. Cambridge: Westview Press, 2003.

Smith, R. *The Utility of Force: The Art of War in the Modern World*. London: Allen Lane, 2005.

United States. *The U.S. Army and Marine Corps Counterinsurgency Field Manual*. Chicago: The University of Chicago Press, 2007.

APPENDIX A

FIRST CANADIAN FORCES' MUSLIM CHAPLAIN DEPLOYED IN AFGHANISTAN[1]

Dr. Daniel Lagacé-Roy

In 2006, Imam Suleyman Demiray, a Sunni Muslim cleric and Padre in the Canadian Forces, was deployed to the southeast of Afghanistan as part of the support of the large numbers of Canadian troops concentrated principally in the Kandahar City region, not a far distance from the Afghanistan/Pakistan border. Padre Demiray is a moderate Imam, committed to ecumenism in its broadest sense. Afghanistan is a Muslim nation struggling to extricate itself from the radical Islamic influence emanating from neighbouring Pakistan.[2] The difficulty in bringing Mullahs of different groups into contact with one another stems predominantly from ethnic, religious and political elements, all of which are infused with Islamic expressions. In other words, to establish dialogue between these groups, Imam Demiray had to foster an openness to relationship that would transcend the entrenched rivalries between the local moderate and pro-government Mullahs and their extremist Taliban counterparts.

The CIMIC (civil-military co-operation) cell within the Provincial Reconstruction Team (PRT), and within which Imam Demiray was serving, has a mandate to assist and effect the restructuring of the Afghanistan National Police (ANP) in Kandahar city. In June 2006, Imam Demiray had the opportunity to experience what it means to get in touch with the *other(s)* by performing his role as a CF Muslim chaplain. With designated funding, twelve 2006 Toyota (extended cab) trucks had been purchased for the local ANP constabulary. These vehicles were modified for police usage and presented to the Chief of Police for Kandahar Province and his local counterpart as a gesture of support for the fledgling ANP in one aspect of a broader program of capacity building. The Acting Commandant of the PRT, Major Eric Liebert, invited Imam Demiray to offer prayer at the opening of the ceremony. Approaching the podium in CF arid pattern (desert) uniform and wearing the traditional Islamic *Takke* (headdress), he began his prayers with an Arabic recitation and concluded with a prayer in both Arabic and English. The effect upon the Muslim audience was immediate: individuals among this predominantly Sunni Pashtun assembly began tugging on their own clothing, excitedly exclaiming in hushed tones to their neighbours that the Imam praying to Allah before them was in a CF uniform. With eyes flashing back and forth between Imam Demiray and their friends, they continually gestured to each other their obvious enchantment. Bringing his prayer to a close, Imam Demiray used the traditional Islamic hand gesture to the face (ears, eyes, and mouth) with the *Amen*. The assembly immediately responded in kind. At the conclusion of the ceremony, the Afghan people in attendance, including dignitaries, queued to hug and shake hands with this Imam in Canadian uniform.

The Pashtun[3] audience at the CIMIC ceremony was confronted with an image clashing paradigmatically with their limited understanding of the *other*. Before this Muslim audience stood Canadian military and civilian police officers with headdress removed, a Christian symbol of reverence and respect for the prayers of this Islamic cleric. An Imam embedded within the Canadian contingent was not opposing religious traditions but was embracing cultural differences. The public appearance of this Muslim cleric in uniform before an Islamic assembly aids in dispelling the long-held belief that Christian and Muslim are foes on the world stage and is a snapshot of the *pluralism* of the Canadian cultural mosaic. Indeed, this case demonstrates that new solutions to persisting problems may be found in ways that respect the dignity of the *other*.

NOTES

1 This case study was informed by the work of Major S.K. Moore and Captain Imam Suleyman Demiray of the Canadian Forces Chaplain Branch; see Moore, S.K. and Suleyman Demiray, "The Canadian Forces Chaplain Branch: Modelling Interfaith Cooperation and Pluralism in Afghanistan," *Ecumenism*, 165 (2007), 5-8.

2 Larry P. Goodson, *Afghanistan's Endless War: State Failure, Regional Politics, and the Rise of the Taliban* (Seattle and London: University of Washington Press, 2001), p.76.

3 Pashtun is the largest ethnic group in Afghanistan comprising 44% of the population, predominantly in the south and is relentlessly subjected to the ubiquitous Taliban propaganda that suggests that, having arrived, the Westerners are intent on changing their way of life. The majority of the Taliban are Pashtun, a shared ethnicity with neighbouring population of Southeastern Pakistan. <www.dfait-maeci.gc.ca/asia/main/country/afghanistan_factsheet-en.asp>, accessed 24 July 2006.

APPENDIX B

CANADIAN FORCES UNIFICATION: A HALF-CENTURY CLASH OF CULTURES

Dr. Robert W. Walker

The concept of integration and unification of the Canadian navy, army, and air services first arose in the aftermath of the First World War. Much later, Minister of National Defence (MND) Paul Hellyer released the *White Paper on Defence* (1964) that reflected his perception that the army, navy and air force were all preparing for different kinds of war. Hellyer wanted a single national strategy, a single coherent defence policy with a single war plan, and a unified system of command. He had no time for the mutually exclusive cultures of the various components of the Canadian military. He also had no single template to apply, as other nations with intra-military cultural differences had structured their military forces in numerous ways, with varying results.[1] Perhaps, had Hellyer anticipated the vernacular of the 21st century, he would have prescribed some early professional development of CQ to "grease the skids" of his unfolding plans.

Cultural conflict with respect to forces unification had existed at government-of-the-day political levels, among deputy ministers and other public service executives, at top navy, army and air force levels, among uniformed members and civilian bureaucrats within DND, and among combat arms regiments, flying squadrons and bi-coastal navies. Round-table discussions aimed at reducing cultural misunderstandings and differences, and at neutralizing suspicions and distrust, appeared to be non-existent. (CQ, in its present form, would not exist until after military leadership and professionalism had been redefined into 21st century versions.) With reference to records of the unfolding events of CF unification, the conflicts in military-civilian relations, and the positions taken by government and military leaders, Major-General Daniel Gosselin recently opined that the situation that commenced then, with variations, persists to this day:

> On one side is the idealistic and progressive concept of unification, the establishment of a single military service in Canada, while on the other side is the traditionalist strong-service idea, focused on the preservation of the army, navy and air force as separate institutions.[2]

So, despite the 45-year evolution of Canada's military forces toward a balanced-force approach, support for distinct and enduring identities and cultures of the three separate services survived. Such loyalties are seen as appropriate and motivating for members. At junior rank levels, this smaller-group loyalty is seen as a component of cohesion, trust, communication and professionalism in fighting units. Nonetheless, these unit loyalties need to be appropriately balanced; that is, unit loyalty should not be so pervasive that loyalty to the CF as a unified institution is weakened or undermined.[3] As Gosselin wrote, the diminishing tug of war between CF unification and the separate navy, army and air force components was an important aspect of the 2005-2008 transformation of the CF:

> ... the CF institution appears well poised to enter the next phase of its evolution... the enduring concept of unification... is returning as the more dominant idea, with the strong-service idea becoming more submissive to the higher needs of Canada's national defence.[4]

The "bottom line" in this case study is: attempts to identify the aspects of cultural differences, recognize the consequences of such diverse perspectives, and implement processes to neutralize undesirable consequences, were never initiated. Those cultural circumstances that were impediments to CF unification and persist even today could have been ameliorated by an implementation of CQ PD, perhaps a timely, focused, 45-year-long version of such.

NOTES

1 J.P.Y.D. Gosselin, "A 50-Year Tug of War of Concepts at the Crossroads: Unification and the Strong-Service Idea". In Allan English, Daniel Gosselin, Howard Coombs, and Laurence M. Hickey (eds.), *The Operational Art: Canadian Perspectives, Context and Concepts*. (Kingston, ON: Canadian Defence Academy Press, 2005) 129-200; J.P.Y.D. Gosselin, "The Storm over Unification of the Armed Forces: A Crisis of Canadian Civil-Military Relations." In Howard G. Coombs (ed.), *The Insubordinate and the Noncompliant: Case Studies of Canadian Mutiny and Disobedience, 1920 to Present*. (Toronto, ON: The Dundurn Group, and Kingston, ON: Canadian Defence Academy Press, 2007) 309-343; and Douglas L. Bland, *Chiefs of Defence: Government and the United Command of the Canadian Armed Forces*. (Toronto, ON: The Canadian Institute of Strategic Studies, 1995). These sources provided detailed information for use in this case study.

2 J.P.Y.D. Gosselin, "A 50-Year Tug of War of Concepts at the Crossroads: Unification and the Strong-Service Idea".

3 See the four recent CF manuals: *Leadership in the Canadian Forces: Leading People* (2007), *Leadership in the Canadian Forces: Leading the Institution* (2007), *Leadership in the Canadian Forces: Conceptual Foundations* (2005) and *Leadership in the Canadian Forces: Doctrine* (2005), (Ottawa, ON: Department of National Defence).

4 J.P.Y.D. Gosselin, "A 50-Year Tug of War of Concepts at the Crossroads: Unification and the Strong-Service Idea," 186-187.

APPENDIX C

REFLECTING ON OKA: THE LEGACY OF MISSION SUCCESS

Denise Kerr and Karen D. Davis

Canadians in general, and the CF in particular, have had, until recently, very little knowledge or understanding of the varied and diverse Canadian Aboriginal cultures, including their contribution to the founding of Canada and its military history. Nowhere was this more evident than on the Kanehsatake Reserve, located in the town of Oka, Québec, where the Iroquoian people were engaged in a centuries-old dispute over their reserve land that in 1774 had spanned 154 square miles but had by 1990 been reduced to 3.6 square miles.

This particular iteration of the land dispute was triggered by the town of Oka's desire to build a golf course on part of the remaining reserve land, specifically, land that held a traditional Iroquoian burial site located in the ancient White Pine Forest. The Aboriginal people were pitted against three levels of government, the police, local citizens, and, later, the CF, in a bitter and violent standoff that lasted 78 days and became known to all Canadians and internationally as the "Oka Crisis". In the end, the CF operation, Op Salon, was deemed an unqualified success by military observers. However, the outcome was dire for the Sûreté du Québec (SQ) and surprising to the CF in the final days of the conflict in particular.

In the months leading up to the crisis, women had demonstrated peacefully in the Kanehsatake forest, and they were present at the barriers at all times throughout the crisis. Yet, neither the SQ nor the CF seemed to be aware of the respective roles of women and men within Mohawk culture and thus the extent to which women were integrated into all aspects of Mohawk activities, including tribal leadership. The SQ officers who approached the barricade in the moments leading up to an exchange of gunfire that resulted in the shooting death of SQ Corporal Marcel Lemay, refused to speak to the Mohawk women, including spokesperson Ellen Gabriel. Insistent upon speaking to a Mohawk leader, the SQ did not understand that there was no single leader. When a dialogue is required to settle a dispute, the women of the Mohawk community are key participants in decision-making, and community decisions are presented by a spokesperson. Upon the request of the Mohawk women, the community faith keeper went to speak to the SQ and relayed the SQ ultimatum back to the community: leave the barrier so the barricades could be removed or the SQ would open fire.[1] When the women responded that they would not be leaving, the SQ stormed the barricades with assault rifles and concussion grenades. In the confusion that followed, each side claimed the other shot first.[2] Over the coming weeks, the situation deteriorated, and eventually the CF was called in by the Premier of Québec as an aid to the civil power.

In several ways, the CF approach to the situation at Oka was exemplary. CF officers on the ground were ordered to walk to the barricade, identify themselves, and shake hands with their counterparts at the barriers on a daily basis. On the first day, there was great concern that people would be killed or seriously injured. One officer laid down his weapon and radios, went over to the barrier, and shook hands. This was a critical point that initiated a dialogue that was sustained for most of the confrontation. The non-commissioned members, for the most part, were also able to demonstrate the depth of their professionalism by staying calm and not reacting to some members of the Mohawk community who, on several occasions, confronted them in a highly aggressive manner.[3]

Importantly, the military went into the Oka situation with a strategy. The Chief of the Defence Staff, General John de Chastelain, knew that the situation and mission at hand represented a potential public relations disaster for the CF. He immediately adopted two fundamental principles: the CF would only use minimum force and the military would not use weapons or attack the Mohawk barricades unless the Warriors took the first shot. In addition, leadership at the policy and strategic level recognized that there would be a battle on the ground for positive publicity, and thus public sympathy. The CF would have to display superior professionalism.[4] In addition, the military carefully adopted discourse and practices that would neither legitimize the Mohawk's claims nor erode public confidence in the CF.[5] As a result, the positive actions of the CF were often highlighted by the media, due at least in part to the "military's successful application of a consistent, coherent and "peaceful" message," throughout the crisis.[6] From the beginning of their involvement, the CF adopted strategy that informed all levels of the operation and subsequently led to mission success in many aspects of the operation.

In spite of the insightful strategy and the professionalism of the CF, which reflected a significant degree of cultural intelligence, Oka presented an incredible challenge for the CF within the broader cultural context of Canadian society. In the short term, the mission was accomplished; however, in the longer term, cultural memory is often refocused on what went wrong and the errors in judgement that influenced the outcomes. For example, the CF and its supporters, for the most part, believed that because Mohawk leadership was not visible and centralized, it was, therefore, disjointed and ineffective. However, centralized leadership is a condition that was placed on negotiations without awareness of how decisions were made within the Mohawk community. The Mohawk women will not forget the disrespect of government and military officials who would not acknowledge them sitting across the table during discussions.[7] Many Canadians will not forget the media footage from the end of the standoff of "a Canadian soldier pushing a Mohawk child to the pavement and then kicking her mother down on top of her...crying children being dragged back and forth after their screaming mothers..."[8] In the case of Oka, the complexities of mission success demanded a well-developed capacity to apply cultural intelligence across all levels of operations – policy, strategic, operational, and tactical – to ensure continued public confidence in the CF.

POSTSCRIPT

Since the events at Oka, much work has been done within the CF to mend the rifts caused by the events of that summer and increase the overall cultural intelligence capacity of the CF. The CF has made many efforts to enhance its awareness of Aboriginal cultures, including hiring an Aboriginal Desk Officer to provide advice to senior officials on the management of Aboriginal affairs within the department, providing Aboriginal people with direct access to an executive level "champion", and the creation of a military-civilian employee Aboriginal advisory group. It has also established an Aboriginal Forum to consult various Aboriginal groups on best practices for the CF with respect to the recruiting and retention of Aboriginals, and has developed several programs such as *Raven* and *Bold Eagle*, and a Pre-Recruit Training Program for Aboriginal youth. In 2007, the Canadian Defence Academy began developing the Aboriginal Leadership Opportunity Year at the Royal Military College (RMC) in Kingston, to help recruit Aboriginal men and women into the officer corps. In 2008, this initiative began its first year of operation at RMC. A similar initiative is being planned for RMC St-Jean in 2009.

NOTES

1 Donna Kahenrakwas Goodleaf, "'Under Military Occupation' Indigenous Women, State Violence and Community Resistance," p.225- 242 in *And Still We Rise: Feminist Political Mobilizing in Contemporary Canada*. (Linda Carty, ed.) (Toronto, ON: Women's Press, 1993), 228-229.

2 P. Whitney Lackenbauer, "Carrying the Burden of Peace: The Mohawks, the Canadian Forces, and the Oka Crisis," *Journal of Military and Strategic Studies*, 10 (2, 2008), 16.

3 Denise Kerr interview with Lieutenant-Colonel Christian Mercier, Commander, CF Language and Recruit School, St-Jean, QC, 2 April 2007.

4 P. Whitney Lackenbauer, "Carrying the Burden of Peace", 27-28.

5 Ibid., 27.

6 Ibid., 43.

7 Donna Kahenrakwas Goodleaf, "'Under Military Occupation'", 231.

8 Jack Todd cited in P. Whitney Lackenbauer, "Carrying the Burden of Peace," 41.

BIBLIOGRAPHY

Alberts, David S. and Richard E. Hayes. *Power to the Edge: Command...Control... in the Information Age*. Washington, D. C.: Department of Defense Command and Control Research Program, 2003.

Arkoun, Mohammed. "For a Subversive Genesis of Values," in *The Future of Values: 21st Century Talks*. Jérôme Bindé (ed.). Paris: UNESCO Publishings, 2004.

Baskerville, Rachel F. "Hofstede never studied culture," *Accounting, Organizations and Society*, 28 (2003).

Bellet, Maurice. *Le Point critique*. Paris: Desclée de Brouwer, 1970.

Bennett, Jennifer J., "Effective Professional Development Strategies for Institutional Leaders" in *Institutional Leadership in the Canadian Forces: Contemporary Issues*. Robert W. Walker, ed. Kingston, ON: Canadian Defence Academy Press, 2007.

Canadian Forces Leaership Institute Project Team. *Broadsword or Rapier? The Canadian Forces' Involvement in 21st Century Coalition Operations*, Kingston, ON: Canadian Forces Leadership Institute, 2008.

________________. *Professional Ideology and the Profession of Arms in Canada*. Toronto, ON: Canadian Institute of Strategic Studies, 2005.

Bland, Douglas L., *Chiefs of Defence: Government and the United Command of the Canadian Armed Forces*. Toronto, ON: The Canadian Institute of Strategic Studies, 1995.

Canada, *Canadian Forces Operations [B-GJ-005-300/FP-000]*. Ottawa, ON: Canadian Forces, 2005.

________. *CDS [Chief of the Defence Staff] Guidance to Commanding Officers*. Ottawa, ON: National Defence, promulgated since 2003.

________. *Gender-Based Analysis: A guide for policy-making*. Ottawa, ON: Status of Women Canada, 1996.

________. *Joint Intelligence Doctrine [B-GL-005-200/FP-000]*. Ottawa, ON: Canadian Forces, 2003.

________. *The Canadian Forces Operational Planning Process [B-GJ-005-500/FP-000]*. Ottawa, ON: Canadian Forces, 2008.

_______. *Land Operations 2021, Adaptive Dispersed Operations: The Force Employment Concept for Canada's Army of Tomorrow*. Kingson, Ontario: National Defence, Directorate of Land Concepts and Design, 2007.

_______. *Leadership in the Canadian Forces: Conceptual Foundations*. Kingston, ON: Canadian Defence Academy, Canadian Forces Leadership Institute, 2005.

_______. *Leadership in the Canadian Forces: Leading People*. Kingston: ON, Canadian Defence Academy, Canadian Forces Leadership Institute, 2007.

_______. *Leadership in the Canadian Forces: Leading the Institution*. Kingston, ON: Canadian Defence Academy, Canadian Forces Leadership Institute, 2007.

_______. *One Team, Many Faces: Employment Equity and Diversity Leadership*, (based on the book *Employment Equity and Diversity Management – A Tip Book* by Lauren Nancoo). Ottawa, ON: National Defence, 2003.

_______. *The Profession of Arms in Canada*. Kingston, ON: Canadian Defence Academy, Canadian Forces Leadership Institute, 2003.

_______. *Canada School of Public Service: Your Partner in Learning*, 2006. Brochure available at www.mySCHOOL-monECOLE.gc.ca.

Challans, Timothy L. *Awakening Warrior: Revolution in the Ethics of Warfare*. New York: State University of New York Press, 2007.

Checkland, Peter. *Soft Systems Methodology in Action*. New York: John Wiley & Sons, 1999.

____________. *Systems Thinking, Systems Practice*. Chichester: John Wiley, 1999.

Clarke, George Elliott. "Contesting a Model Blackness: A Meditation on African-Canadian Americanism, or the Structures of African Canadianite," *Essays in Canadian Writing*. 63 (1998), 1-52.

Clausewitz, Carl von. *On War*, Michael Howard and Peter Paret (eds.), (Oxford: Oxford University Press, 1976).

Coles, John P. "Incorporating Cultural Intelligence into Joint Doctrine," *Iosphere: Joint Information Operations Center*. (Spring 2006).

Cook, Martin. *The Moral Warrior*. New York: State University of New York, 2004.

Curtis, Liane, Dipti Gupta and Will Straw. *Culture and Identity: Ideas and Overviews, paper commissioned by the Department of Canadian Heritage*. Ottawa, ON: Department of Canadian Heritage, 2001.

Davis, Karen D. "Women, Gender, and Cultural Intelligence in the Canadian Forces," paper presented at *Defending Democracy: Diversity in the Security Sector* conference, Queen's University, Kingston, ON, November 1-3, 2007.

____________, "Gender," in *The Military Leadership Handbook*. Bernd Horn and Robert W. Walker, eds. Toronto and Kingston: Dundurn Press and Canadian Defence Academy Press, 2008

____________ and Brian McKee. "Culture in the Canadian Forces: Issues and Challenges for Institutional Leaders" in *Institutional Leadership in the Canadian Forces: Contemporary Issues*. Robert W. Walker, ed. Kingston, ON: Canadian Defence Academy Press, 2007.

Durivage, Andre and Jacques Barrette. "The Role of Values and Organizational Culture in a Strategic Planning Process," in *Strategic Planning for Police*. Dan Ogle, ed. Ottawa, ON: Canadian Police College, 1991.

Earley, Christopher and Randell S. Peterson. "The Elusive Cultural Chameleon: Cultural Intelligence as a New Approach to Intercultural Training for the Global Manager," *Academy of Management Learning and Education*, 31 (1, 2004), 106.

____________ and Soon Ang, eds. *Cultural Intelligence: Individual Interactions Across Cultures*. Stanford, California: Stanford University Press, 2003.

Edwards, Commander R. S. *Leadership in Canadian Forces Operations: Emerging Issues and Future Trends*. Kingston, ON: Canadian Forces Leadership Institute, 2007.

Elliott, Anthony. *Concepts of the Self*. Cambridge, UK: Polity Press, 2001.

English, Allan D. *Understanding Military Culture: A Canadian Perspective*. Montreal & Kingston: McGill-Queen's University Press, 2004.

English, John A. *Lament for an Army: The Decline of Canadian Military Professionalism*. Toronto, ON: Irwin Publishers, 1998.

Ford, Kimberly-Anne and Karen D. Davis. *Cultural Intelligence, Emotional Intelligence and Canadian Forces Leader Development: Concepts, Relationships and Measures*. Kingston, ON: Canadian Forces Leadership Institute and Defence Research and Development Canada, 2007.

Forsythe, G. B., S. Snook, P. Lewis and P. Bartone. "Making Sense of Officership: Developing a Professional Identity for 21st Century Army Officers," in L. J. Matthew, ed. *The Future of the Army Profession*. New York: McGraw-Hill Primis Custom Publishing, 2002.

Foucault, Michel. "The Ethics of the Concern of the Self as a Practice of Freedom," in *Michel Foucault: Ethics, Subjectivity and Truth*, Volume 1. Paul Rabinow, ed. New York: The New Press, 1997.

Geertz, Clifford. *The Integration of Cultures*. New York: Basic Books, 1973.

Giber, David, Louis Carter and Marshall Goldsmith. *Best Practices in Leadership Development: Handbook*. San Francisco, CA: Linkage Inc. and Jossey-Bass / Pfeiffer, 2000.

Goodleaf, Donna Kahenrakwas. ""Under Military Occupation" Indigenous Women, State Violence and Community Resistance," in Linda Carty, ed. *And Still We Rise: Feminist Political Mobilizing in Contemporary Canada*. Toronto, ON: Women's Press, 1993.

Goodson, Larry P. *Afghanistan's Endless War: State Failure, Regional Politics, and the Rise of the Taliban*. Seattle and London: University of Washington Press, 2001.

Gosselin, J. P. Y. Daniel. "The Storm over Unification of the Armed Forces: A Crisis of Canadian Civil-Military Relations," in Howard G. Coombs, ed. *The Insubordinate and the Noncompliant: Case Studies of Canadian Mutiny and Disobedience, 1920 to Present*. Toronto, ON: The Dundurn Group and Kingston, ON: Canadian Defence Academy Press, 2007.

____________________. "A 50-Year Tug of War of Concepts at the Crossroads: Unification and the Strong-Service Idea". In Allan English, Daniel Gosselin, Howard Coombs, and Laurence M. Hickey (eds.), *The Operational Art: Canadian Perspectives, Context and Concepts*. Kingston, ON: Canadian Defence Academy Press, 2005.

Gray, Colin S. *Another Bloody Century: Future Warfare in the 21st Century*. Oxford: Oxford University Press, 2005.

___________. *Modern Strategy*. London: Weidenfeld, 1999.

Guillebaud, Jean-Claude. *Le goût de l'avenir*. France: Éditions du Seuil, 2003.

Gutting, Gary, ed. *The Cambridge Companion to Michel Foucault*. 2nd edition. New York, NY: Cambridge University Press, 2005.

Hammes, T. *The Sling and the Stone: On War in the 21st Century*. Osceola, WI: Zenith Press, 2004.

Hiemstra, R. *Lifelong Learning: An Exploration of Adult and Continuing Education with a Setting of Lifelong Learning Needs*, 3rd Edition. Fayetteville, NY: Hitree Press, 2002.

Hofstede, G. "Images of Europe," *The Netherlands Journal of Social Sciences*. 30 (1994), 63-82.

Lee, Chay Hoon amd Klaus J. Templer. "Cultural Intelligence Assessment and Measurement," in *Cultural Intelligence: Individual Interactions Across Cultures*. C. P. Earley and S. Ang, eds. Stanford, California: Stanford University Press, 2003.

Horn, Bernd and Bill Bentley, "The Road to Transformation: Ascending from the Decade of Darkness." in *Institutional Leadership in the Canadian Forces: Contemporary Issues*. Robert W. Walker, ed. Kingston, ON: Canadian Defence Academy Press, 2007.

Huges, Richard L. and Katherine C. Beatty, *Becoming a Strategic Leader*. San Francisco, CA: Jossey-Bass & the Center for Creative Leadership, 2005.

Inglehart, Ronald and Pippa Norris. *Rising Tide: Gender Equality and Cultural Change Around the World*. Cambridge University Press, 2003.

______________ and Christian Welzel. *Modernization, Cultural Change, and Democracy: The Human Development Sequence*. Cambridge University Press, 2005.

______________, Miguel Basañez and Alejandro Moreno, *Human Values and Beliefs: A Cross-Cultural Sourcebook*. Ann Arbor: The University of Michigan Press, 2005.

Jary, David & Julia Jary. *Web-linked Dictionary: Sociology*. Glasgow: Harper Collins Publishers, 2000.

Jeffery, Michael K. *Inside Canadian Forces Transformation: Institutional Leadership as a Catalyst for Change*, Kingston, ON: Canadian Defence Academy Press, Forthcoming.

Johnson, James P., Tomasz Lenartowicz, and Salvador Apud. "Cross-cultural Competence in International Business: Toward a Definition and a Model," *Journal of International Business Studies* 37(4, 2006).

Johnston, Alastair. "Thinking About Strategic Culture" *International Security* 19 (4, 1994-1995).

Kealey, Daniel J., David Protheroe, Douglas MacDonald and Thomas Vulpe. *Readying your Organization for International Projects*. Ottawa, ON: Department of Foreign Affairs and International Trade, Canadian Foreign Service Institute, Centre for Intercultural Learning, 2004.

Kegan, Robert. *The Evolving Self: Problems and Process in Human Development*. Cambridge: Harvard University Press, 1982.

___________. *In Over Our Heads: The Mental Demands of Modern Life*. Cambridge: Harvard Press, 1994.

Kirke, Charles. "The Organizational Cultural Approach to Leadership: 'Social Structures' – A Tool for Analysis and a Way Ahead," p. 288-291 in *Dimensions of Military Leadership*. Allister MacIntyre and Karen D. Davis (eds.). Kingston, ON: Canadian Defence Academy Press, 2006.

Kristeva, Julia. *Powers of Horror: An Essay on Abjection*, translated by L.S. Roudiez. New York: Columbia University Press, 1982.

Lackenbauer, P. Whitney, "Carrying the Burden of Peace: The Mohawks, the Canadian Forces, and the Oka Crisis," *Journal of Military and Strategic Studies*, 10 (2, 2008), 1-71.

Lagacé-Roy, Daniel. "Institutional Leader Ethics" in *Institutional Leadership in the Canadian Forces: Contemporary Issues*. Robert W. Walker, ed. Kingston, ON: Canadian Defence Academy Press, 2007.

____________ and Janine Knackstedt. *Mentoring Handbook*. Ottawa, ON: Department of National Defence, 2007.

Lawrence, Thomas B. and Roy Suddaby. "Institutions and Institutional Work," in *The Sage Handbook of Organization Studies*. 2nd Edition. Stewart R. Clegg, Cynthia Hardy, Thomas B. Lawrence and Walter R. Nord (eds.). London: Sage Publications, 2006.

Lynn, John. *Battle: Combat and Culture*. Boulder: Westview Press, 2003.

Maffesoli, Michel. "Towards a Postmodern Ethic of the Aesthetic," in *The Future of Values: 21st Century Talks*. Jérôme Bindé, ed. Paris: UNESCO Publishings, 2004.

Maguire, Steve, Bill McKelvey, Laurent Mirabeau and Nail Öztas. "Complexity Science and Organizational Studies," *The Sage Handbook of Organization Studies*. 2nd Edition. Stewart R. Clegg, Cynthia Hardy, Thomas B. Lawrence and Walter R. Nord, eds. London: Sage Publications, 2006.

McCarthy-Senebald, Maryanne. *The Development of Expertise in Maritime Surface and Subsurface (MARS) Officer Training in the Canadian Navy*. Calgary, AB: University of Calgary, 2000. Unpublished Ph.D. dissertation.

McCauley, Cynthia D, Russ S. Moxley and Ellen Van Velsor. eds. *Handbook of Leadership Development*. San Francisco, CA: Jossey-Bass & the Center for Creative Leadership, 1998.

Mills, C. Wright. *The Sociological Imagination*. Oxford: Oxford University Press, 1959.

Milne, G. *Making Policy: A Guide to the Federal Government Policy Process*. Ottawa: Queens Printer, 2004.

Moore, S.K. and Suleyman Demiray. "The Canadian Forces Chaplain Branch: Modelling Interfaith Cooperation and Pluralism in Afghanistan," *Ecumenism*, 165 (2007), 5-8.

O'Toole, Roger. "Canadian Religion: Heritage and Project," in D. Lyon and M. Van Die, (eds.) *Rethinking Church, State, and Modernity: Canada Between Europe and America*. Toronto: University of Toronto Press, 2000.

Oxford Dictionary of Philosophy. Oxford: Oxford University Press, 1994.

Pike, H. R.N. MacLennan, and N.M. Perron. *Canadian Forces Diversity Climate Project: 1999 Survey Administration*. Ottawa, ON: Department of National Defence, Director Human Resources Research and Evaluation, 2000.

Rabinow, Paul. *Michel Foucault: Ethics, Subjectivity and Truth*. Volume 1. New York: The New York Press, 1997.

Rennick, Joanne Benham. *Religion in the Ranks: Religion and Military Culture in the 21st Century*. Draft Doctoral Thesis. Waterloo, ON: University of Waterloo, 2008.

Robbins, Stephen P. and Nancy Langton, *Organizational Behaviour: Concepts, Controversies, Applications*. 3rd Canadian Edition. Toronto: Prentice Hall, 2003.

Schein, Edgar. *Organizational Culture and Leadership*. San Francisco: Jossey-Bass, 1992.

Selmeski, Brian R. *Military Cross-Cultural Competence: Core Concepts and Individual Development*. Kingston, ON: Royal Military College of Canada, Centre for Security, Armed Forces & Society, 2007.

Senge, Peter. *The Fifth Discipline: The Art and Practice of the Learning Organization*. New York: Currency/Doubleday, 1990.

Simons, Rupert. *Professional Military Learning: Next Generation Professional Military Education in the New Zealand Defence Force*. Canberra, Australia: Air Power Development Centre, 2005.

Smith, Rupert. *The Utility of Force: The Art of War in the Modern World*. London: Allen Lane, 2005.

Spencer, Emily. *Crucible of Success: Applying the Four CQ Domain Paradigm. The CF in Afghanistan as a Case Study* (Technical Report 2007-05). Kingston, ON: Canadian Forces Leadership Institute, 2007.

Stacey, Ralph D. *Managing the Unknowable: Strategic Boundaries Between Order and Chaos in Organizations*. San Francisco: Jossey-Bass, 1992.

Tamas, Andy. *Warriors and Nation Builders: Development and the Military in Afghanistan*. Kingston, ON: Canadian Defence Academy Press, 2009.

Taylor, Carl. "The ACIDD Test: a Framework for Policy Planning and Decision-Making" *Optimum, The Journal of Public Sector Management*, 27 (4, 1997).

Taylor, Charles. *A Secular Age*. Cambridge: The Belknap Press of Harvard University Press, 2007.

___________. *Les sources du moi. La formation de l'identité moderne*. Paris: Éditions du Seuil, 1998.

Thomas, David C. "Domain and Development of Cultural Intelligence: the Importance of Mindfulness." *Group and Organisation Management* 31 (2006).

___________ and Kerr Inkson. *Cultural Intelligence: People Skills for Global Business*. San Francisco: Berrett-Koehler Publishers, Inc., 2003.

Turner, Bryan S. "Cosmopolitan Virtue, Globalization and Patriotism." *Theory, Culture & Society*. 19, (1, 2002), 45-63.

Ulin, Robert C. "Revisiting cultural relativism: old prospects for a new cultural critique," *Anthropological Quarterly*. 80 (3, Summer, 2007).

United States. *Mixed Blessings: US Government Engagement with Religion in Conflict-Prone Settings.* Washington: Center for Strategic and International Studies, 2007.

__________. *The U.S Army and Marine Corps Counterinsurgency Field Manual.* (Chicago: The University of Chicago Press, 2007).

Urban, S. and I. Goldenberg. *Canadian Forces Diversity Climate Project 2005 Administration.* Ottawa, ON: Department of National Defence, Director Personnel Applied Research, 2006. Sponsor Research Report 2006-08.

Vance, C. M. *Analysis of Self-initiating Career Path Strategies: Similarities and Differences across Genders.* Paper presented at Annual Meeting of the Academy of Management, Seattle, WA, 2002.

Vulpe, Thomas, Daniel Kealey, David Protheroe and Doug MacDonald. *The Profile of the Interculturally Effective Person.* Ottawa, ON: Centre for Intercultural Learning, Canadian Foreign Service Institute, Foreign Affairs Canada, Second Edition, 2001.

Vuola, Elina."Remaking Universals?: Transnational Feminism(s) Challenging Fundamentalist Ecumenism." *Theory, Culture & Society.* Vol. 19, No. 1, 2002, 175-195.

Walker, R. W. ed. *Institutional Leadership in the Canadian Forces: Contemporary Issues*, Kingston: Canadian Defence Academy Press, 2007.

__________. *The Professional Development Framework: Generating Effectiveness in Canadian Forces Leadership*," Kingston, ON: Canadian Forces Leadership Institute, 2006, Technical Report 2006-01.

Williamson, Murray and Mark Grimsley. "Introduction: On War," in *The Making of Strategy: Rulers, States, and Wars*, Williamson Murray, Alvin Bernstein, and MacGregor Knox, eds. Cambridge: Cambridge University Press, 1996.

Winslow, Donna. *The Canadian Airborne Regiment in Somalia: A Socio-cultural Inquiry.* Ottawa, ON: Ministry of Public Works and Government Services, 1997.

____________, Lindy Heinecken, and Joseph L. Soeters. "Diversity in the Armed Forces," *Handbook of the Sociology of the Military*. Giuseppe Caforio, ed. New York: Kluwer Academic/Plenum Publishers, 2003.

Wong, Leonard, Stephen Gerras, William Kidd, Robert Pricone and Richard Swengros. *Strategic Leadership Competencies*. Carlisle Barracks, PA: United States Army War College, Strategic Studies Institute, 2003.

Wright, Ronald. *A Short History of Progress*. Toronto, ON: House of Anansi Press Inc., 2004.

CONTRIBUTORS

Major Brent Beardsley has been a Staff Officer in the Leadership Lessons Learned Section at the Canadian Forces Leadership Institute since 2002. The focus of his research and work is on officer leadership issues. Major Beardsley is an Infantry Officer with 30 years of service with the Royal Canadian Regiment including four tours of regimental duty, a wide range of staff appointments on extra-regimental employment and service on operations. He holds a Pre-Arts Diploma from Sir George Williams University, a Bachelor of Arts degree in History from Concordia University, a Post-Graduate Diploma in Education from McGill University and a Masters of Applied Science in Management from the Royal Military College of Canada. He is currently completing his second Masters degree in War Studies at RMC where the focus of his studies is on genocide and humanitarian intervention in the current and future security environment.

Dr. Bill Bentley is a retired infantry officer currently serving in the Canadian Forces Leadership Institute as a Section Head. He is the author of numerous articles and research essays. His book *Professional Ideology and the Profession of Arms* in Canada was published in 2005. Dr. Bentley received the Deputy Minister/Chief of the Defence Staff (DM/CDS) Innovation Award in 2005 and was awarded the Meritorious Service Medal in 2006 for his work on the reform of the Canadian Forces Professional Development System.

Lieutenant-Commander (retired) **Karen D. Davis** is a defence scientist with Defence Research and Development Canada, currently employed as a Team Leader and Section Head at the Canadian Forces Leadership Institute, holds a Master of Arts in Sociology from McGill University and is a Ph.D. candidate at the Royal Military College of Canada. Karen is a contributing writer to *Leadership in the Canadian Forces: Leading the Institution* (2007), and is the editor and a contributing writer to, *Women and Leadership in the Canadian Forces: Perspectives and Experience* (2008).

Denise Kerr has been a member of the Public Service of Canada for over 32 years, and has worked for various government departments, including the Correctional Service, Privy Council Office and National Defence. She holds a Bachelor of Military Arts and Science from the Royal Military College of Canada and currently works as Senior Staff Officer for Governance and Policy Coordination at the Canadian Defence Academy.

Dr. Daniel Lagacé-Roy is a professor at the Royal Military College of Canada in Kingston where he is presently teaching Military Professionalism and Ethics, and Psychology and Philosophy of Religious Conflicts amongst other courses. He is also an associate researcher in Military Ethics and Leadership at the Canadian Forces Leadership Institute. He previously taught Ethics at the

Dufferin-Peel Catholic School Board in Mississauga, Ontario, at the Université du Québec in Rimouski, Quebec and at the University of Alberta in Edmonton, Alberta. He served in the Canadian Forces from 1987 to 1995 (Regular) and from 1998 to 2001 (Reserves). He recently published *Ethics in the Canadian Forces: Making Tough Choices* (workbook and instructor manual) and a *Mentoring Handbook*. Dr. Lagacé-Roy received his Ph.D. in Philosophy from Université de Montréal, a Master of Arts in Theology and Philosophy, a Teaching degree from L'Université Laval, Quebec and a second Master of Arts in Counseling from St-Paul University in Ottawa.

Dr. Robert W. Walker, a graduate of the Royal Military College of Canada and Queen's University, served for 21 years in the Royal Canadian Navy and the Canadian Forces, including an assistant professorship at RMC. He next served in the Royal Canadian Mounted Police for 17 years as a research psychologist and civilian member of the Force. Since 2002, Dr. Walker has been an Associate Professor at the Canadian Forces Leadership Institute. Recently, he contributed numerous chapters to, and co-edited, *The Military Leadership Handbook*, a 2008 commercial publication by Dundurn Press, Toronto and Canadian Defence Academy Press, Kingston.

Justin C. Wright is a graduate of St. Thomas University and the University of New Brunswick in Fredericton. He holds a Bachelor of Arts and a Master of Arts degree in Sociology specializing in addiction studies and social theory. He is a defence scientist with Defence Research and Development Canada, currently assigned to the Canadian Forces Leadership Institute of the Canadian Defence Academy in Kingston, Ontario. His research interests include identity development, military socialization and cross-cultural competency.

ACRONYMS AND ABBREVIATIONS

ACIDD	Analysis, Choice, Implementation, Debate and Decision
ADM (Pol)	Assistant Deputy Minister – Policy
AI	area of interest
ANP	Afghanistan National Police
AO	area of operations
CANCOM	Canada Command
CDS	Chief of the Defence Staff
CEFCOM	Canadian Expeditionary Force Command
CF	Canadian Forces
CFLI	Canadian Forces Leadership Institute
CIDA	Canadian International Development Agency
CIMIC	civil-military co-operation
CQ	cultural intelligence
DFAIT	Department of Foreign Affairs and International Trade (Canada)
DLCD	Directorate of Land Concepts and Design
DND	Department of National Defence
EQ	Emotional Intelligence
FAC	Foreign Affairs Canada
4GW	4th Generation Warfare
IPB	intelligence preparation of the battlespace
JIMP	Joint, inter-agency, multinational and public domains
MND	Minister of National Defence
NATO	North Atlantic Treaty Organisation
NCM	non-commissioned member
NGO	non-governmental organization
NORAD	North American Aerospace Defence Command
OGD	other government department
OPP	operational planning process
OR	operational research
PD	professional development
PDF	professional development framework

PRT	Provincial Reconstruction Team
RAND	Research and development (name of U.S. based non-profit, non-partisan research and analysis institution)
RCMP	Royal Canadian Mounted Police
RMA	Revolution of Military Affairs
RMC	Royal Military College of Canada
SAT	Strategic Advisory Team
SJS	Strategic Joint Staff
SOD	systemic operational design
SQ	Sûreté du Québec
SSM	soft systems methodology
3D	defence, diplomacy and development
UK	United Kingdom
UN	United Nations
UNEF	United Nations Emergency Force
UNFICYP	United Nations Force in Cyprus
US	United States (of America)
WoG	whole of government
WVS	world values survey
WWI	World War I
WWII	World War II

GLOSSARY

4th Generation Warfare (4GW) Refers to combat characterized by a blurring of the lines between war and politics, soldier and civilian, peace and conflict, battlefield and safety.

Andragogy The theory of teaching adults. First put forward by German teacher Alexander Kapp in 1833 as *Andragogik* to describe elements of Plato's education theory and use of Socratic teaching/learning methods incorporating questions and discussions for adult students; characterized as experience-based, learner-centred, facilitator-guided, and student-active methods of teaching.

Attitude Learnt and enduring tendency to perceive or act towards persons or situations in a particular way.

Belief A firm opinion or conviction; an acceptance of a thing, fact, statement, etc.

Belief System The configuration of beliefs that exist in a particular society or culture.

Command The authority vested in an individual of the armed forces for the direction, coordination, and control of military forces. Also, the authority-based process of planning, organizing, leading, and controlling the efforts of subordinates and the use of other military resources to achieve military goals.

Complex System An open system operating under conditions that are far from equilibrium, and comprised of a large number of parts, each of which behaves according to some rule or force that relates it interactively to other parts.

Cross-Cultural Competence The ability to function effectively in other than one's own culture as a result of personal attributes, knowledge, and skills.

Cross-Cultural Savvy	The ability to understand cultures beyond one's organizational, economic, religious, societal, geographical, and political boundaries, as well as anticipate and understand the values, assumptions, and norms of other groups, organizations, and nations.
Cultural Awareness	A combination of culture-specific and culture-general knowledge.
Cultural Intelligence (CQ)	The ability to recognize the shared beliefs, values, attitudes and behaviours of a group of people and, the capacity to effectively apply this knowledge toward a specific goal or range of activities.
Cultural Intelligence (military)	The analysis of social, political, economic, and other demographic information that provides understanding of a people or nation's history, institutions, psychology, beliefs, and behaviours.
Cultural Literacy	The ability to understand one's own individual patterns and one's own cultural norms.
Cultural Relativism	The belief that human values, far from being universal, vary a great deal according to different cultural perspectives.
Culture	The shared beliefs, values, attitudes and behaviours of a group of people.
Culture-General Knowledge	Information that addresses how to understand cultural differences, knowing one's own culture and how it affects the mission, cultural norms and values, and skills for bridging cultural differences.
Culture-Specific Knowledge	Information and descriptions, often factual and targeting tactical-operational level missions, including, for example, history of local area and roots of conflict, basic components

	of culture (e.g. values, traditions), language training, physical geography and climate, and economic patterns.
Emotional Intelligence	The ability to monitor one's own and others' emotions, to discriminate among them, and to use the information to guide one's thinking and actions.
Ethics	Sub-field of philosophy that aims at clarifying and evaluating the nature of right and wrong. It involves the recognition and promotion of certain values, the development of ethical principals and obligations, which can be used to help guide behaviours and deal with ethical issues, and the development of moral reasoning in all ranks of military professionalism so that military members can understand and reflect their own actions.
Ethnocentrism	The belief that one's own group is the most important or culturally superior to other groups; often characterized by an incapacity to acknowledge that cultural differentiation does not imply the inferiority of ethnic groups to one's own. Institutional ethnocentrism refers to the imposition of home culture ways of doing things without considering differences in other cultures.
Ethos	The essential and distinctive character or spirit of a race, or people, or a system, culture, institution… defined by customs, usages, habits, morals.
Hermeneutics	The art, skill, theory, and philosophy of interpretation and understanding, predominantly in relation to the reading of texts, and also of the whole social, historical and psychological world.
Identity	Sense of self, of personhood, constructed in relation to sameness and difference from

others and derived from individual characteristics such as gender or ethnicity, for example, that can be relatively durable, as well as identities developed through a process of "construction" influenced by cultural characteristics such as norms, beliefs, values, traditions, organization, sub-groups which may be relatively fluid over a lifetime.

Institution	A formally established organization with a specific function to perform consistently over time; an institution has legal or quasi-legal standing and permanence. Institutions dependent upon bureaucracy and hierarchy are generally slow to adapt and change, while networked institutions organized around professional ideals will be relatively more adaptable to changing circumstances.
Knowledge Management	An integrated systematic approach which when applied to an organization enables the optimal use of timely, accurate and relevant information; it also facilitates knowledge discovery and innovation, fosters the development of a learning organization and enhances understanding by integrating all sources of information, as well as individual and collective knowledge and experience.
Leadership	The process of directly or indirectly influencing others, by means of formal authority or personal attributes, to act in accordance with one's intent or a shared purpose; *effective* leadership in the CF is defined as directing, motivating and enabling others to accomplish the mission professionally and ethically, while developing or improving capabilities that contribute to mission success.
Learning organization	An organization that is able, on an ongoing basis, to critically examine its performance, assimilate information from the environment, and transform itself, with a view to adapting

to challenges and positioning itself to exploit opportunities or to establish a dominant capability.

Meaning-Making Process whereby individuals construct mental models that ground their understanding of the social world, as well as their own status within it.

Military Ethos The living spirit that creates and shapes military culture, finds full expression through the conduct of members of the profession of arms, and is comprised of three fundamental components: beliefs and expectations about military service, Canadian values, and Canadian military values.

Military Strategy The bridge that cements military power to political purposes, comprised of numerous dimensions related to people and politics, preparation for conflict, and war and conflict.

Moral Reasoning The means by which individuals recognize and resolve ethical difficulties; informed by personal construction of the social world.

Operational Level The level of activity acting as a bridge between strategy and tactics; 'operational art' includes coordination of diverse assets, both military and non-military, and the fusion of the abstract and general strategic objectives and the somewhat mechanical, technical activities at the tactical level into a functional formula referred to as the campaign plan.

Organization A specific grouping of entities connected by function or interest and with an identifiable end state or purpose. Conventionally, organizations, public or private, take on specific structures with hierarchical or networked characteristics focused on the production of a given product, material, service or intellectual. Organization may be short-term or long-term; permanent organizations develop history and culture and may be very similar to, or become, an institution.

Pedagogy — The function of teachers, incorporating lecturer-centred teaching of dependent students, typically children, who are considered to be a relatively passive audience or a classroom of students in receipt of "expert" presentations.

Phenomenology — A set of theoretical approaches that attempt to understand the way in which people experience the world they create and inhabit; the study of human experience and consciousness in everyday life which emphasizes the intentionality of consciousness, the immediacy of individual experience.

Professional Ideology — A framework guiding the profession of arms that claims specialist, theory-based knowledge and a commitment to ethical values that guide the application of that knowledge.

Professionalism — In general, displaying the qualities or features of a profession. With respect to the CF, professionalism means that CF members apply their unique body of military expertise in accordance with the civic, legal, ethical, and military values of the military ethos, pursuant to the profession's responsibility to society and a strong personal identification with military activities and the military way of life.

Social Intelligence — The ability to understand the feelings, thoughts, and behaviours of persons, including oneself, in interpersonal situations and to act appropriately on that understanding.

Societal Imperative — With respect to the CF's legal and professional responsibilities to society, the general obligation to serve Canada's interests and satisfy Canadians' expectations of the armed forces, including specific obligations to remain subordinate to civil authority and to reflect core Canadian values.

Society	A society can be defined as a population, whose members are subject to the same political authority, occupy a common territory, have a common culture and share a sense of identity.
Soft Systems Methodology	Multi-iterative model (originally seven stages) used to explore organizational problem contexts with an emphasis on learning, appreciative understanding of and recognition of different and competing worldviews; most recent focus on learning and process, rather than formal systems theory.
Strategic/Strategy	See military strategy.
Strategic Corporal	Information age, post-modern phenomenon by which the individual soldier, sailor or air person can be responsible for making a decision and taking an action which can have positive or negative consequences on the outcome of an operation at tactical, operational and strategic levels, and within national and international contexts.
Strategic Culture	The socially transmitted habits of mind, tradition and preferred methods of operations that are more or less specific to a particular geographically based security community.
Systems Thinking	A discipline with various frameworks, spanning the physical and social sciences, engineering and management, for seeing wholes, interrelationships, and patterns of change.
Tactical Level	The level at which tangible activity is delivered to satisfy the context set by strategy and operational art. The factors of fear, danger, fatigue and extreme physical and mental exertion are either unique to the tactical realm or their effects are greatly magnified there.

Transformation (Canadian Forces)	The program of fundamental and real change, initiated in 2005 by the Chief of Defence Staff, to generate a renewed military institution better able to meet CF needs and CF mission effectiveness through vision, strategy, strategic planning, restructuring, and reconstituting CF purpose for the 21st century.
Values	Beliefs concerning what is centrally important in life and what should, therefore, guide decisions and actions; properties or qualities that make something useful, desired, or esteemed; result from evaluations and a concept of the desirable.

INDEX